Culture and Structure at a Military Charter School

New Frontiers in Education, Culture, and Politics

Edited by Kenneth J. Saltman

New Frontiers focuses on both topical educational issues and highly original works of educational policy and theory that are critical, publicly engaged, and interdisciplinary, drawing on contemporary philosophy and social theory. The books in the series aim to push the bounds of academic and public educational discourse while remaining largely accessible to an educated reading public. *New Frontiers* aims to contribute to thinking beyond the increasingly unified view of public education for narrow economic ends (economic mobility for the individual and global economic competition for the society) and in terms of efficacious delivery of education as akin to a consumable commodity. Books in the series provide both innovative and original criticism and offer visions for imagining educational theory, policy, and practice for radically different, egalitarian, and just social transformation.

Published by Palgrave Macmillan:

Education in the Age of Biocapitalism: Optimizing Educational Life for a Flat World
    By Clayton Pierce

Schooling in the Age of Austerity: Urban Education and the Struggle for Democratic Life
    By Alexander J. Means

Critical Pedagogy and Global Literature: Worldly Teaching
    Edited by Masood Ashraf Raja, Hillary Stringer, and Zach VandeZande

Culture and Structure at a Military Charter School: From School Ground to Battle Ground
    By Brooke Johnson

Culture and Structure at a Military Charter School

From School Ground to Battle Ground

Brooke Johnson

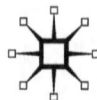

CULTURE AND STRUCTURE AT A MILITARY CHARTER SCHOOL
Copyright © Brooke Johnson, 2014.

All rights reserved.

An earlier version of Chapter 5 was originally published by Men & Masculinities. The original article can be found at http://jmm.sagepub.com/content/12/5/575.full.pdf+html.

First published in 2014 by
PALGRAVE MACMILLAN®
in the United States—a division of St. Martin's Press LLC,
175 Fifth Avenue, New York, NY 10010.

Where this book is distributed in the UK, Europe and the rest of the world, this is by Palgrave Macmillan, a division of Macmillan Publishers Limited, registered in England, company number 785998, of Houndmills, Basingstoke, Hampshire RG21 6XS.

Palgrave Macmillan is the global academic imprint of the above companies and has companies and representatives throughout the world.

Palgrave® and Macmillan® are registered trademarks in the United States, the United Kingdom, Europe and other countries.

ISBN: 978–1–137–36092–2

Library of Congress Cataloging-in-Publication Data is available from the Library of Congress.

A catalogue record of the book is available from the British Library.

Design by Newgen Knowledge Works (P) Ltd., Chennai, India.

First edition: October 2014

10 9 8 7 6 5 4 3 2 1

Contents

List of Tables	vii
Acknowledgments	ix
1 Introduction	1
2 Schools in the Crosshairs: Neoliberalism, Militarization, and Public Education	35
3 Sending Good Kids to Military School: Why Parents Choose the MEI?	59
4 Reading, Writing, Arithmetic, and War: Militarized Pedagogy and Militarized Futures	83
5 A Few Good Boys: Masculinity at the MEI	97
6 Ask, Tell, Talk Back: Queering Resistance to Gendered Heteronormativity	115
7 Conclusion	139
Notes	157
References	161
Index	183

Tables

1.1	Enrollment by race	24
3.1	Teacher Credential Rates and Experience	66
3.2	CAHSEE Passing Rates for MEI, District, and State	67
3.3	CST Scores for English Language Arts and Math	67

Acknowledgments

This project could not have taken place without the gracious allowance of Major Allen West who took a risk and allowed a curious researcher to explore the fascinating school and students he so very much believed in. Of course this project could also not have come to fruition without the students, faculty, parents, and the staff at the Military Educational Institute. I thank them for sharing their lives with me and allowing me to bother them with questions, tag along to school events, and take part in their school and community. I would particularly like to express gratitude to Adalberto Aguirre, Jr, at the University of California, Riverside, for his support from the beginning to the end of this project and his valuable feedback, patience, and calm demeanor throughout it all. I would be remiss not to mention Linda Kim at Arizona State University who read multiple drafts and somehow always managed to provide encouragement at the most needed moment, and David Romero, a student at Northeast Illinois University, who generously helped with editing and formatting. Finally, I am particularly indebted to my colleagues at Northeastern Illinois University who read drafts, offered advice, support and much needed motivation. A heartfelt thank you to Laurie Fuller, Christina Gomez, Erica R. Meiners, and Brett Stockdill.

Introduction

It is almost 8 a.m.; cadets are picking their way across the dew-covered grass toward the back courtyard for morning formation. Backpacks and lunchboxes are discarded as cadets step off the grass onto the cement square to shuffle toward their platoons where they stand huddled together, sleepily awaiting commands. Teachers mill around the courtyard with clipboards, checking uniforms, taking note of who has arrived while keeping the more spirited students in line. Several cadets carrying four dark brown show rifles arrive with the US and California flags. The rifles are dispersed among the color guard at the top of the courtyard. A young campus aide, Sarah, struggles to pull the mobile speaker system across the field despite its small plastic wheels catching in the wet grass. A few mothers stand on the grass near the color guard, discussing school events and swapping information about the cheapest place to buy the required school uniforms. Their arms folded as they keep one eye trained on their child. The school's commandant strides across the field wearing a bright yellow polo with the school's insignia embroidered in black on the left breast, pressed black pants, and shined shoes. As he approaches, the students gather together; platoon leaders shout commands that ring out across the courtyard. The group of huddled students snaps to attention in perfectly aligned blocks, staring ahead, unflinching. The commandant nods to the color guard and to Sarah. The flags are presented, rifles slide through gloved hands, and a solemn military march sputters from the depths of the mobile speaker system.

US culture is filled with the images, values, and overtones of the military: surveillance, cell phone tapping, live television broadcasts of invasions and bombings, bumper stickers that read "Support the Troops" and "Got War?," metal detectors, bullet-shaped cologne dispensers, boot camp–style fitness classes, the National Rifle Association, Troops-to-Teacher programs, camouflage-clad brides, air shows, urban warfare video games, and the War on Poverty, Crime, Drugs, and Global Terrorism. Our public spaces are being booby-trapped, and our private lives bombed by a culture of violence and war inline with military ideologies and values. Schools such as Virginia Tech and Sandy Hook Elementary have become the public spaces for the enactment and media spectacle of societal violence in the form of mass shootings and nonstop media coverage. As we lay prone in front of our TV sets and watch the latest invasion unfold and the latest scenes of domestic and global violence, neoliberalism churns our public lives into private supersized profits.

The connection between militarization and neoliberalism is not a recent phenomenon as the US military has long been the trailblazer of US neoliberal policies, deploying the US military in defense of US global financial and corporate interests. However, the trend toward the privatization and corporatization of public education and the implementation of neoliberal values into the public education system is heralded as a quick and easy fix for low test scores, underperformance, and violent city schools. The privatization and corporatization of public education runs parallel with the increasing militarization of the education of US children, where metal detectors, bag searches, security officers, uniforms, and military personnel are standard in US public schools. Local schools are being transformed into publicly funded, privatized militarized zones with no one questioning the rapid transformation from public to private, from civilian to enlisted, and from learning ground to battleground. The militarization and invasion of a culture of militarism into public education could not have occurred without the simultaneous advancement of neoliberal policies and ideals.

Neoliberalism is a political economic theory premised upon the argument that free markets, free trade, and private property rights are avenues to human prosperity and freedom. One outcome of neoliberal educational reforms is publicly funded charter schools. The first charter school in the United States opened its doors in 1992. As of the 2011–2012 school year (most recent data available at the time of publication), there were 5696 charter schools nationwide (US Department of Education 2013b), with the state of California leading with 985 charter schools followed by Texas with 581 and Arizona with 531 (US Department of Education 2013c). Charter schools are exempt from some of the curricular and structural requirements of regular public schools and an emerging trend among charter schools across the United States is the increasing number of militarized charter schools—the Military Educational Institute (MEI) is one such school. How did the US public education system get to this point, where militarized educational institutions are viewed as a solution to failing US schools? What effects will schools like the MEI have on communities and students they serve?

This book highlights the nexus of neoliberal school policies and the militarization of US public education through a qualitative study of a fully militarized charter school in Southern California. Specifically, this book examines militarization and the interactions of race, class, gender, and sexualities in the lives of students enrolled at the MEI, their parents, and teachers. The questions addressed in this book are: How does a militarized school affect student lives and their families? How do school practices at a militarized charter school promote neoliberal goals and values? How do school personnel and students promote the mission of a militarized charter school? Does militarization affect the construction of gender and sexualities at the school? Finally, is there resistance to militarization, and if so how is it enacted? To explore these connections, I utilize theories of education, sociological theories on gender, and cultural studies including discourse analysis and queer theory. I specifically utilize queer theory and cultural studies to extend the theoretical framework of reproduction and resistance theories of education. I conclude with ways in which to discursively and materially resist militarization of youth across personal, community, institutional, and global levels.

In order to fully grasp the everyday lives and choices of the cadets and parents at the MEI, it is important, I argue, to have a complete understanding of the intricacies between neoliberalism, militarization, and public education. Thus, in this chapter I set the theoretical context for the remainder of the book. I begin with a brief overview of militarization and how it differs from militarism. A section in which I outline neoliberal ideology and its application to public education follows. I argue that neoliberal reforms exploit and exacerbate social inequality and are concentrated in poor communities and schools of color coinciding with the militarization of public education. This book not only examines the everyday culture and practices of militarization at the MEI but also how MEI students resist and challenge these practices. To that end, a overview of theories of education, including reproduction and resistance theories, precedes a general overview of the organization of this book.

Militarism, Militarization, and Education

Enloe (2000, 3) defines militarization as the "step-by-step process by which a person or a thing gradually comes to be controlled by the military or comes to depend for its well-being on militaristic ideas." Militarization is both a discursive and a material process involving the ongoing process of aligning cultural, institutional, and economic forms with militaristic values, beliefs, and practices. What is particularly powerful about militarization is its subtlety in that as institutions, practices, and people become militarized, the transformation and outcomes are viewed as acceptable, valuable, and, more importantly, normal (Enloe 2000). For example, in a militarized society or community, it becomes acceptable and thus normalized that the effective solution to any number of social issues, such as failing pubic schools or inner city crime, is militarized discipline such as on-campus surveillance video and city police raiding homes with flash grenades. Even the media is militarized as war and imagery of war are considered a genre of entertainment and highly profitable (Mann 1992).

Although militarism and militarization are often used interchangeably, they are quite separate terms with quite disparate theoretical explanatory power. Militarism is a much narrower concept referring to a cultural project in which ideologies and priorities are inline with martial values. Militarism often resides in the political realm and refers to moments of heightened military frenzy as in times of war preparation. Militarization, on the other hand, "draws attention to the simultaneously material and discursive nature of military dominance" (Lutz 2002, 725) to shape society. Thus, militarization makes sense of how institutions and practices are defined and valued though the dominance of the military and ongoing and subtle process of societal transformation. The importance of militarism should not be dismissed, however, as a culture of militarism allows for or assists with the acceptance and naturalization of militarization (Adelman 2003).

When examining militarization and its impacts and effects on social institutions such as education, it is important to employ an intersectional lens. Doing so illustrates how the process and effects of militarization vary across and within groups as well as how militarization intersects, interacts, and fuels inequalities, privileges, and power. Social structures of inequalities as well as racist, sexist, and homophobic ideologies can be employed to further militarization and violence. For example, Saltman and Gabbard (2003, 2011) examine how public schools are sites for the enforcement of neoliberal ideologies and corporatization of education through the strong arm of militarization. In *Sexual Decoys: Gender, Race and War*, Eisenstein (2007) illuminates how the militarization of gender, race, sexuality, and feminist principles further US neoliberal imperial politics. Smith (2005) examines the link between militarization and environmental racism highlighting the extensive and almost exclusive practice of nuclear testing on indigenous lands. Finally, Stephen (2008) contends that the militarization of the US-Mexico border has had deadly and gendered consequences on those who live and cross through these border zones. Taking a critical and intersectional approach to understanding militarization illustrates how power is wielded across institutions (like public education) and structures of inequalities and fans the flames of continued militarization. In the following sections, I will examine the application of neoliberal ideology to public education across two broad neoliberal ideological concepts: choice and accountability.

Neoliberalism: Roots and Emergence in US Education

Public education reform, since the 1980s, is predicated upon neoliberal economic doctrine and includes trends such as militarization, privatization, and corporatization of US schools (Giroux 2009; Robbins 2008; Saltman and Gabbard 2011). Coupled with neoliberal ideals and policies, these trends are presented as solutions for a failing US public education system. Most of these failing schools are underfunded, overcrowded, and located in poor and working-class communities of color (Saltman 2000) and few students who attend these schools "make it." It is the failing or crisis of public education in the United States that creates the allowance for neoliberal reform, which usually takes place as privatization of public education in the form of vouchers, charter schools, and corporate control and oversight of testing.

Milton Friedman, the ringmaster of free-market capitalism, argued that moments of crises—actual or perceived—are the only times when significant change can occur (1962, ix). In *The Shock Doctrine: The Rise of Disaster Capitalism*, Naomi Klein (2007) examines how neoliberalism rose to prominence through the exploitation of disaster. That is, when societies are experiencing shock due to a disaster (wars, natural disasters, economic crashes), the societal disorientation can be exploited to impose neoliberal economic doctrines. This is what Klein (2007) terms "disaster capitalism" or the corporate reform of societies in shock, and this began with the reforms and privatization that took place with the Pinochet regime (based on Friedman's advisement) in the mid-1970s in Chile. While Klein's work looks specifically at the neoliberal aftermath and greed of actual disasters, such the 2004 tsunami in the Indian Ocean, Friedman himself argued that there only has to be a perceived crisis in order to implement neoliberal reforms. Low reading scores of city schools (Banchero 2010b), below-average competency in math of US students since 2000 (Layton 2013), and the raced and classed portrayals of violent poor schools (Giroux 2009; Robbins 2008) are examples of such an opportunity of crisis in public education.

6 CULTURE AND STRUCTURE AT A MILITARY CHARTER SCHOOL

Similarly to Klein, Harvey (2005) describes it as "accumulation by dispossession," and Saltman (2007) argues that current privatization initiatives and neoliberal ideology in education reform are part of "smash and grab" capitalism that aims to destroy public schools in order to privatize them and must be understood as the intersection of neoliberalism with neoconservative ideology. Neoliberal ideology contends that public control of public resources (education, health care, social security) is inefficient and overly bureaucratic, while private control is efficient through the discipline of the market. As many scholars have pointed out (Klein 2007, Saltman 2007), there is no starker example of accumulation by dispossession than New Orleans post–Hurricane Katrina. Local and national business interests took advantage of the chaotic aftermath of Hurricane Katrina to outright plunder and privatize the city's public housing and education system. For example, just weeks after the hurricane the state dismissed all public school teachers, took over a hundred public schools, and began turning them over to private organizations (Saltman 2007). Prior to the hurricane, there were five charter schools, and in the 2012–2013 school year 72 of 90 schools in New Orleans were charter schools (Cowen Institute 2013).

An important and disturbing aspect of neoliberal logic is that capitalism is equated with democracy. Within neoliberal ideology, democracy is defined not through a political or social context but rather as the unchaining of markets from state control. Democracy under neoliberalism is an economic form of corporate control, increasing consumerism, the ability to accumulate profit, and the turning over of public holdings to private interests. The role of the state is to ensure that the right conditions are in place to sustain such practices—to ensure economic freedom. Thus, freedom is defined as freedom from the oversight and control of the state and has led to the dismantling of the welfare state. The laws and health of the marketplace take precedence over the public good, and the social state, through neoliberal values and goals, is transformed into a corporate state, serving the interests of the marketplace. Thus, public resources are converted into resources for corporate gain and control through privatization schemes. This is particularly true for public education as public schools are increasingly defined as a source of private profit (rather than a public good) through privatization schemes such as vouchers and corporate-run charter schools.

Although proponents argue that neoliberalism is democratic, it is directly opposed to democracy and democratic ideals. Neoliberalism exacerbates social inequalities (Navarro 2006), as social inequalities and social problems are economically defined and constructed as another market for profit accumulation (the privatization of prisons is a great example see Selman and Leighton 2010). Additionally, those citizens who cannot or do not consume (such as those victimized by historical social processes of racism, sexism, homophobia, or inequalities of class) are in need of control by a disciplinary and surveilling state. In *Discipline and Punish*, Foucault (1977) sets forth that the modern society (modern neoliberal state) controls people through the imposition of norms (normalization). The ability or inability of people to meet normative standards is used to justify reforms or "disciplinary" measures of control. The "disciplinary" aspect of modern society, inherent in neoliberal theory, is pervasive and extends to a variety of aspects of social life and social institutions. For education this translates to the discipline of accountability measures such as standardized testing and merit pay for teachers, as well as disciplinary policies of zero tolerance and militarization of public schools to control "violent" raced and classed students.

Chubb and Moe (1990) recommended a new system for US public education in their book, *Politics, Markets and America's Schools*. Their proposal was heavily influenced by Friedman and based educational reform and success upon free markets and four related concepts of neoliberal ideology: competition, choice, accountability, and efficiency. They argue that market choice–based education allows the clientele (parents and students) the power to switch from school to school in order to find a school that best meets their needs and expectations. Schools are, then, in competition with one another and are accountable to please parents and students, as parents and students always have the *choice* to switch schools. Schools are also accountable to the state in regard to minimum educational criteria, nondiscrimination laws, standardized test scores, and "anything else that would promote informed choice among parents and students" (Chub and Moe 1990, 224–225). According to the theory, schools that are unable to satisfy a large enough population or market share will fail and close. The schools that survive, however, do so because they are not only able to effectively compete for the business of parents and students, but they are also successful by performing as efficiently as possible and stripping away the "dysfunctions" of bureaucracy (Chubb and Moe 1990). According to Chubb and Moe, US education as a whole will improve as noncompetitive (read "bad") schools will fail, while competitive (read "good") schools will prosper. This will simultaneously raise the educational performance and standards of US students. One of the most successful neoliberal "choice" reforms, is charter schools and they are examined in the following section.

Choice and Charter Schools

Charter schools are public, as they are funded with state tax monies, but are exempt from most state laws governing school districts.[1] Charter schools are an alternative to traditional public schools and viewed as a way out of failing schools. Additionally, charter schools are unique in that they are not held to many state controls (teacher unions, teacher credentials, curriculum, for example) and are allowed to build curriculum and structure the schools under a variety of models including but not limited to science, art, college prep, vocational, and militarized.

Although charter schools have been quite popular across the United States, they are not without drawbacks. First, charter schools drain off already scarce district and statewide resources. There is limited money available for financing public education, and charter schools redirect portions of this money away, thus limiting the finances available for struggling schools. Additionally, most charter schools locate in areas where schools are failing because competition from these schools is low (Henig and MacDonald 2002; Morest 2002), and because charter schools are funded on a per-pupil basis, funding is diverted away from those underfunded schools that are already failing. In addition, charter schools spend more on administrative costs and less per pupil than noncharter schools (Miron and Urschel 2010). It must also be kept in mind that many charter schools are often run by for-profit organizations. Thus, not all the funding a charter school receives is put toward educational and student needs as profits remain an important part of the organizational framework.

Second, charter schools are headed by nonelected boards and governing bodies that are filled with people with business or corporate backgrounds. This results in the fundamental dismantling of democratically elected school boards and councils and moves the school's oversight away from communities into the private sector (Lipman and Hirsch 2007; Saltman 2007). Thus, charter schools work under the antidemocratic, neoliberal assumption that public institutions are inefficient and in need of business discipline.

Third, there is increasing evidence that charter schools intensify race and class segregation (Cobb and Glass 1999; Garcia 2008; Miron and Urschel 2010; Morest 2002; Yancey 2000). The National Study of Charter Schools, based on annual phone surveys, field visits, student achievement tests, teacher surveys, and analysis of charter school laws, rulings, procedures, policies, and legislation (Nelson et al. 2000), found that charter schools are serving predominantly at-risk students and students of color. Manno et al.'s (1999) study of charter schools in Washington, DC, found that charter schools in the city enroll a disproportionate number of Black and Latino students. Additionally, Henig and MacDonald (2002) found that charter schools are more likely to locate in areas with high proportions of Black and Latino residents. The most recent Stanford CREDO (2013) study, which includes charter data from 27 states, found that 54 percent of all charter students live in poverty.

While attracting and enrolling at-risk students and students of color is not in and of itself negative, studies have shown that charter schools perform on par or even lower than noncharter public schools (Banchero 2010a; CREDO 2013; Loveless 2003), and the achievement gap holds when controlling for differences of race and class (Nelson et al. 2004). Additionally, Fusarelli (2002) found that charter schools in Texas, while enrolling a disproportionate number of students of color, actually underenrolled special education and limited English proficiency students. This suggests that charter schools may be "creaming" among students of color (Morest 2002, 43). Additionally, charter schools rely on new and inexperienced teachers, and turnover is high (Brown and Gutstein 2009; Darling-Hammond 2010). Cheap and inexperienced teachers who burnout quickly teach students who already receive the least in terms of educational resources but arguably need the most. Thus, charter schools are proposed as a solution to failing public schools, but similar to other public choice programs, charter schools aggravate the problem of race and class segregation which they are argued to assuage and instead provide needy students with the least.

Lastly, as mentioned earlier, charter schools are free to construct curriculum and manage schools in ways that are somewhat independent of state or district control. One of the most popular school organizational styles is the corporate model in which for-profit companies manage charter schools. The argument follows that charter schools will be innovative and efficient when managed my corporations, but most corporate-run charter schools have fallen back on traditional education models and rote learning (Saltman 2012). In addition, many charter schools rely on additional nonstate funding by way of donations, such as those from the Gates Foundation. The reliance on philanthropic funds further narrows the education models charter schools utilize (Saltman 2010). Militarized charter schools, such as the MEI, are also becoming increasingly popular (Banchero and Sadovi 2007; Wedekind 2005) and provide the touted neoliberal discipline to fix failing schools and control "dangerous" youth. As charter schools are more likely to be amid darker and poorer communities, it comes as no surprise that militarized charter schools would enter the neoliberal fray.

Neoliberal theorists like Freidman, Chubb, and Moe greatly influenced public education in the United States, particularly in reinforcing the idea of individual choice through the free market. Education is no longer viewed as a public good, but is rather constructed as a private good based on individual choice and competition. Choice rhetoric and reforms do not take into account the historical and structural social inequalities and differing levels of social power (Ladson-Billings 2006) and operate under several assumptions. First, such arguments assume that schools have equal resources and are able to compete against one another equally. However, resources vary greatly between states and school districts (Kozol 1991, 2005; Moser and Ross 2002). The inequality between schools and districts makes fair competition impossible and poorer, less-funded school districts are unable to compete with richer, better-funded schools. Schools that fail justify neoliberal thinking and policies since they will be forced to improve without additional funding (Levin 1998).

Even if it were possible to have schools competing equally, competition between schools is still a drain on scarce school resources as some resources are shifted away from academics toward marketing and public relations. Additionally, schools try to attract students who will improve their image or reputation (creaming), leaving out many special or high needs students. Thus, a subtle shift has occurred from "student needs to student performance and from what the school does for the student to what the student does for the school" (Apple 2001a, 413). The schools are able to justify this process due to limited space in such schools.

Next, market-based school reform is premised upon the ability of parents and students to shop for the school that best fits their needs. However, this also rests upon several assumptions, the first being that there is a range of schools for parents and students from which to choose. Many rural and suburban areas lack such realistic choices, as the nearest school may be many miles away (Lubienski 2004, 77). Even when choices are available, transportation costs discourage families from switching (Andre-Bechely 2007; Goldhaber et al. 2005). Additionally, it is usually middle-class parents who switch because they are the ones most likely to have the resources, knowledge, skills, and contacts to navigate through the system (Apple 2001a, 415; Goldhaber et al. 2005). One must note that academic reasons are not always central in choosing schools as many parents may switch to a new school due to the proximity of the school from work, home or day care (Andre-Bechely 2007; Goldhaber et al. 2005).

Finally, there is little evidence that good schools are open to or able to accommodate transfers, or that students actually did change schools. For example, in the first year of the No Child Left Behind Act (NCLB), very few students eligible for transfer actually did. In Los Angeles, of 200,000 students eligible for transfer, less than 50 did. Of Chicago's 125,000 students eligible for transfer, less than 800 switched, and in

New York, just 1507 of the 220,000 students eligible for transfer did so (Brownstein 2003, 216).

Choice programs have been widely touted as the solution to the problems in the US education system. The proof, however, has not been in the pudding. The most likely participants in choice programs are privileged groups. When at-risk students and students of color do participate in choice programs, it is often through charter schools as charter schools as mentioned earlier are more likely to locate in areas where existing schools are failing and these schools often have a large number of people of color (CREDO 2013; Henig and MacDonald 2002). Thus, choice programs are more likely to reinforce education inequality rather than improve education quality. Proposals for school choice are an attempt to redefine the boundaries between the public and private sectors by transforming the social good of education into a private enterprise that benefits corporations, private business interests, and those students who are already ahead (Brown 2003; Saltman 2011), as well as offers poor communities fewer resources (Kozol 2005; Leonardo and Grubb 2014). There is no better example than the outright selling off of public education in New Orleans after Hurricane Katrina (discussed previously) (Lipman 2011b).

Accountability and Efficiency, NCLB, and Race to the Top

The accountability movement in education is premised upon the neoliberal assumption that failing schools are inherently inefficient and in need of discipline. It holds schools accountable and operates under the assumption that standardized tests and measurements can and do measure education quality. Standardized testing is not a recent phenomenon. It came into use after World War I to sort children by cognitive ability. The accountability movement of today emerged in 1994 when Congress during the Clinton administration passed a mandate that required all schools to show, through standardized testing, annual student progress toward state-set goals (Peterson 2003). Tests, according to theory, would push schools to become more efficient at educating youth and enforcing improvement through sanctions and penalties for poor test performance. The NCLB and Race to the Top program (RTT) are quintessential examples of neoliberal arguments for accountability.

NCLB was signed into law by President Bush in 2002. NCLB offers "choice" as a three-part plan to failing and underperforming schools as measured by standardized tests. First, NCLB mandates that school districts with schools that fail to meet statewide proficiency goals for two consecutive years must offer public choice options (e.g., charter schools, nonfailing public schools, etc.) to all their students. Second, if schools fail to meet statewide proficiency goals for a third consecutive year, 20 percent of a local school district's Title I funds must be used for public choice. Lastly, any school that fails for a fourth consecutive year will either be taken over by the state government or placed under private management ("Fact Sheet: No Child Left Behind Act" 2002). The push for private takeover is based upon the neoliberal logic that privatization will result in greater efficiency as bureaucracy is reduced. Additionally, all schools must reach the impossible task of 100 percent proficiency by 2014. Thus, NCLB makes schools accountable to the state, the parents, and the students through state-mandated standards and proficiency levels and links good test scores with education quality. However, the unrealistic and unattainable goals set by NCLB results in the privatization of public schools as more and more schools fail and are taken over by private organizations.

NCLB is working under two false assumptions. First is that test scores are a valid indicator of success and failure of schools. Schools may very well be labeled as "failing," but may be very successful in educating large populations of students with diverse educational needs. Thus, many schools will be forced to drop successful policies and programs in preparation for testing (Neill et al. 2004). Second, NCLB assumes that harsh sanctions will lead to school improvement without taking into consideration how that improvement should come about. Additionally, by solely focusing on school failure, social issues that sustain school inequality will fail to be addressed, thereby shifting blame to schools and educators for the consequences of structural inequalities such as classism and racism (Neill et al. 2004, 18).

Like other operationalized concepts of neoliberalism, accountability and NCLB have not gone without scrutiny. One of the criticisms of neoliberal reform is the drain on scarce educational resources. Beside the fact that NCLB was severely underfunded (Neas 2003), the spread of standardized testing of student proficiency funnels money into corporate educational institutions that specialize in testing, curriculum, and supplementary materials and away from other educational needs like teacher training (Bracey 2005). Thus, money flows from the federal government to schools and directly into the hands of for-profit organizations as schools are increasingly relying on marketing and corporate sponsors for school supplies and other programs (Arce et al. 2005; Kozol 2005).

Additionally, because of the severe sanctions of NCLB, schools are narrowing instructions to test material and lower-order thinking (Neill et al. 2004; Volante 2004). Even with increased attention to test preparation, standardized tests are specifically designed to ensure some amount of failure. Additionally, the assumption is that standardized tests measure objective, quantifiable knowledge, but standardized tests have inherent racial, class, and gender biases, making it more difficult for schools in racially diverse communities to reach achievement levels set by NCLB (Crouse and Trusheim 1988; Epps 2002; Fleming and Garcia 1998; Jencks and Philips 1998; Jensen 1980; Taylor 1981, 2002, 2003). In fact, in 2004, the National Center for Fair and Open Testing reported that "no school serving low-income children will pass" (Neil et al. 2004, 3) and that over 70 percent of schools would fail (11). Thus, the drive for accountability and NCLB mistakes measuring school outcomes for school improvement and pushes thousands of children toward failure. Neoliberal reformers and supporters of privatization, however, can use the evidence of "failed" schools to push for further privatization (Karp 2004, 58–59). Labeling a school as "failing" also makes it more difficult to attract and keep qualified and talented teachers (Darling-Hammond 2004, 13). But at the core, labeling a school as failing is not necessarily about improving educational outcomes or teacher effectiveness, but rather about justifying privatization and corporate takeover of public education.

Neoliberal reforms such as privatization and corporate takeover of public schools is not supported by evidence (as outlined in Saltman 2012), but is rather justified by neoliberal ideology and profit-driven greed. Public education is in the eyes of neoliberal reformers and corporations, a frontier of unexploited, publically funded profit. In 2009–2010, the United States spent $638 billion on public elementary and secondary schools (US Department of Education 2013a). For a perspective of what is at stake for profit accumulation, in 2011, venture capital transactions in K-12 education totaled $389 million compared to $13 million in 2005 (Simon 2012). What is more, the label of "failing school" is not applied equally, but is a raced and classed project. For example, in a national-level study of all grade levels across the United States, Logan et al. (2012) found that schools with higher percentage of students of color performed worse. As race and class are interrelated, these findings were correlated with poverty, but they held even when controlling for class.

The Obama administration tried to make changes to NCLB during reauthorization in 2010, but was hampered by US Congress. In 2012 the Obama administration announced a plan to ease some of the restrictions of NCLB through flexibility waivers in proficiency standards in reading and mathematics. In return, schools that receive waivers must set up new standards that prepare students for college. However, the central accountability component of NCLB—reliance on standardized testing to evaluate schools and learning outcomes—still stands.

RTT rests upon similar neoliberal arguments as NCLB that accountability and competition between schools, or between states, leads to improved public education. The Obama administration announced RTT in 2009 and is a contest funded by the US Department of Education that scores states on a point system for awarding federal grant money. Points are awarded for implementing corporate reforms and policies such as standard-based assessment for students and educators, adoption of common core curriculum, and implementation of data systems to track and assess states and educational goals. RTT pushes states to adopt neoliberal reforms by offering up badly needed funds. What is more, RTT works under the assumption that schools are competing for federal grant money on a level-playing field and obscures historical and structural inequalities. Thus, RTT, much like NCLB, exacerbates existing inequalities in public education while simultaneously funneling more funds into privatized hands and students into private schools.

As mentioned earlier the US education system was originally formulated on the principles of equal opportunity and common good (Larabee 1997). But these ideals have been worn away as schools reflect the stratified job market. Although corporate models of education and oversight have always been part of the US education system, it was not until the work by Milton Friedman (1955, 1962) and the 1966 report (*A Nation at Risk*) that neoliberal ideals were exalted as the fix-all for a failing public education system. The initial success of magnet schools in the 1960s, 1970s, and 1980s illustrated the proclaimed power and ability of neoliberalism to fix a broken US education system and ushered in, with the help of Chubb and Moe (1990), an array of educational programs and reforms based upon market competition, choice, efficiency, and accountability.

However, the research reviewed and discussed in this chapter shows that these programs did little to alleviate the inequalities and failing public education system. More often than not, such programs exacerbated existing inequalities, turning the common good of equal opportunity in education into a private good. Education has become, in the logic of the market, something to be consumed, a commodity to be purchased with public tax dollars through school vouchers, magnet schools, charter schools, or standardized tests. Withholding these same tax dollars is an effort to instill discipline through simulating a competitive marketplace that punishes failing schools. The encroachment of neoliberalism upon public education in the past 30 years has gone hand-in-hand with the increasing militarization of public schools, as failing schools and raced and classed youth need not only neoliberal discipline but also militarized discipline. The rise of militarization in public schools is the focus of the following chapter in which I argue that such programs are little more than recruitment ploys for vulnerable youth who are harvested for enlistment through militarized educational programs due to historical, economic, and educational injustices.

Theories of Education

Social reproduction theorists argue that children of the working class overwhelmingly end up in working-class occupations, and that although public education is argued to be the great equalizer, schools perpetuate inequality rather than combat it. Reproduction theorists examine how schools serve the interests of the ruling or dominant class by perpetuating the skills and attitudes that legitimize social structures. Kozol's (2005) book, *The Shame of the Nation*, illustrates in sordid detail how underfunded inner city schools partner with local businesses and corporations in curriculum development and implementation of school-to-work programs to guide even the brightest students into working-class occupations and lives. Henceforward I present a review of theories of education, specifically reproduction theories (both deterministic and cultural models), resistance theories, and their applications, to examine neoliberal reform and militarization of public education.

Reproduction theorists fall into two main camps: deterministic and cultural models. Theorists of deterministic models of reproduction utilize Marxist theory to examine the structural requirements of the capitalist economic system and illustrate how individuals fulfill predetermined class-based roles. Althusser (1969, 1971) argues that schools are the major institution for the reproduction of class formation as schools not only train workers in the necessary skills to fulfill specific social locations but also instill the appropriate attitudes and behaviors that are in line with the division of labor, such as respect for authority and discipline. Schools are, according to Althusser, the dominant institution for not only skill and knowledge acquirement but also, more importantly, ideological control that perpetuates the class structure. Althusser's theory of social reproduction is critiqued for the assumption that schools function in a smooth and conflict-free manner. His theory is overly deterministic, as social actors within this theory do not exhibit agency in interacting with social structures and institutions such as schools. Althusser leaves no room for the interplay between domination and subordination, resistance, or the creation and use of oppositional ideologies by subordinated groups (Hall 1981).

Similarly, Bowles and Gintis (1976) argue that schools train the wealthy and upper class to occupy the top-level jobs, while the poor and working class are prepared for the bottom rungs of society. Schools do this through the reproduction of skills necessary for the hierarchical division of labor, but also through the reproduction of values and dispositions that perpetuate capitalist class formations. Schools, according to Bowles and Gintis, reflect the social division of labor as schools located in poor and working-class communities focus on rules, regimentation, and control of behavior. Schools in wealthier neighborhoods, on the other hand, "favor greater student participation, less direct supervision, more student electives, and, in general, a value system stressing internalized standards of control" (1976, 132–133). Additionally, parents of poor or working-class youth want their children to learn the value of rules and control, as they know from personal experience that this is important to occupational success. Middle-class parents, based on their experiences in the workplace, expect schools to be more open with their children. Again, similarly to Althusser, Bowles and Gintis were critiqued for the overly simplistic and deterministic nature of social reproduction (Giroux 2001). Bowles and Gintis do not account for the role of ideology or consciousness within schools or how social actors interact and mediate the knowledge schools reproduce.

Deterministic theories of social education fail to take into account notions of power and resistance and the interplay between domination, subordination, and resistance in both materialistic and ideological ways. Domination, whether ideological or economic, is never total or complete. What is missing from overly deterministic theories of social reproduction is an account of the role of culture in social reproduction. It is at this point in which cultural models of reproduction step in and examine how cultures work to perpetuate social inequality and include such theorists as Bourdieu and Passerson (1977), Bernstein (1977), and Heath (1983).

Bourdieu and Passerson (1977) move on from this overly deterministic point toward a more nuanced theory of social reproduction that links power with culture. They argue that cultural capital, or the intergenerational experiences, knowledge, and skills of a particular social class, is the key to understanding social reproduction. According to Bourdieu (1979), the dominant class imposes control not through brut economic force but through the imposition of a worldview that aligns with its own interests and is accepted as natural. Bourdieu refers to this more subtle wielding of power as "symbolic violence". Schools are one of the most important methods of symbolic violence and reproduction of social formations of class. Schools simultaneously reward the cultural capital of the dominant class (under the guise of neutrality and objectivity) and shun or dismiss knowledges of subordinated groups such as people of color, women, as well as class and sexual minorities. Thus, children of the dominant class succeed in schools (due to their familiarity with dominant cultural values and skills) and secure superior occupations.

There is a second important piece to Bourdieu's theory of social reproduction that moves beyond cultural power within social structure to social practice and the actual dispositions of social actors. Habitus, or the "subjective but not individual system of internalized structures, schemes of perception, conception, and action common to all members of the same group or class" (Bourdieu 1977, 82–83), is the link between social structure and practice. In other words, an individual's worldview of opportunity is shaped by the internalized, subjective probabilities of those around them and this in turn shapes behavior. For example, children in the lower class will examine the lives and experiences of those around them and will set their ambitions lower than a similar child who was surrounded by successful middle-class individuals. Thus there is a symbiotic relationship between cultural capital (structure) and habitus (practice), where structures produce subjective understandings and aspirations, which in turn reproduce such structures. What is important to note is that since domination is viewed as natural subordinated groups participate in their own subjugation.

Thus, for Bourdieu, social class determines the success or failure in school, albeit in a more nuanced way than laid out by Althusser or Bowles and Gintis. However, Bourdieu's theory is not without weaknesses as it only understands dominant culture to be a form of culture and ignores subjugated cultures that can perpetuate alternative understandings of the social world as well as resistance. That is, subjugated groups can resist dominant institutions and forms of culture across a variety of axes, such as race/ethnicity, gender, sexualities, abilities and feminisms, and create new spaces of oppositional culture and knowledge. Bourdieu, much like other reproduction theorists, views domination as one-sided and top-down without understanding how subjugated groups make sense of social structures and dominant cultural knowledges.

Following the theoretical lines of Bourdieu, Bernstein examines linguistic patterns across social classes in the process of social reproduction. Bernstein (1977) argues that each class has distinct language patterns. Children of the working-class learn, through socialization, "restricted" linguistic codes, while middle-class children are socialized to use "elaborated" codes. Within restricted codes, meanings remain implicit and bound by context. Elaborated codes, on the other hand, are explicit and tied to the unique experience and social location of the speaker. However, schools utilize elaborated codes placing middle- and upper-class kids at an advantage over poor kids and children of the working class.

Similarly, Heath also looks at linguistic codes, but factors in race to the analysis. Through ethnography, Heath (1983) examines the linguistic differences between two working-class schools, one White and one Black, in a rural community in the Southern United States. Heath finds, similarly to Bernstein, that language codes taught at home either hinder or support academic success. Heath found that working-class Black kids' linguistic patterns focused on comparative answers rather than specific information. The lack of language patterns that dealt with specific information led Black students to fail in school as specific information in regard to language patterns was utilized at school. White students fared better as many of them learned these language skills in elementary school, but many still lacked mastery.

Bernstein and Heath examined how language incongruence between schools and home life impedes educational success of lower-class students and students of color. However, in much the same trappings as Bourdieu, Bernstein and Heath fail to recognize the power of agency and resistance to dominant forms of culture. Ultimately, reproduction theorists explain how schools are in line with the needs of the dominant class and social formations, but do little to critique or question the structure and everyday processes and interactions of schooling.

Thus, in some sense, reproduction theorists reproduce the status quo by failing to adequately critique power and domination as well as to recognize that subordinated groups are not homogenous or passively accept dominant material and economic renderings of the social world. Both deterministic and cultural models of social reproduction dismiss schools as sites of agency and power, thereby dismissing the ability of social actors to interact with and resist within schools and the broader community. More recently, resistance theories such as Willis (1977) and Giroux (2001) argued that oppositional cultures both inside and outside of schools are ways in which subordinate groups can and do challenge hegemony.

Reproduction theorists tend to focus on the structural constraints of schools and dismiss the agency of students and how they navigate school processes and everyday interactions of domination as well as resistance. Giroux (1983a), however, argues that resistance to schooling by students is an important area for inquiry that should not be ignored as it offers an avenue to move past the structure-agency conundrum. Resistance theory examines the structural constraints and domination that individuals and groups face while simultaneously highlighting the response to these structures in the form of practices and attitudes, which are cultural as well as political. As MacLeod (1987) explains, "oppositional cultural patterns draw on elements of working class culture in a creative and potentially transformative fashion. Thus, the mechanisms of class domination are neither static nor final" (22).

Critical ethnographic studies have illustrated how resistance to school structures and process is an important part of school life (Everhart 1983; McRobbie 1978; Willis 1977). Willis (1977) examines, through extensive participant observation, the "counter school culture" of White, working-class males in Britain. Those boys or "lads" who adhere to the counterschool culture believe school will not enable social mobility (based on examples from their own families and workplaces) and thus spare no expense to exert their antischool stance. Additionally, the "lads" equate working-class occupations and particularly manual labor with masculinity, and mental labor with femininity. This leads these young boys to actively and willingly choose lower-class occupations and, as Willis argues, to eventually and unwittingly accept class domination. Additionally, Fine's (1991) study found that students who drop out from alternative high schools in the Bronx borough of New York City had a more critical understanding of inequality and injustice than those who remained in school and were more likely to challenge educational norms.

Although resistance theorists have succeeded in highlighting agency and oppositional cultures within schools, they are less adept at highlighting that not all forms of oppositional culture are resistant against subordination and domination. Resistance theories also fail to see how oppositional culture may be a reaction against historical forms of domination (racism, sexism, classism, homophobia) that originate and exist outside of schools (Giroux 2001). We must then examine the expansive social context and everyday practices in which resistance by subordinated groups is played out such as the ongoing processes of neoliberalism, militarization, racial oppression, gender, class, and sexualities injustices.

Another shortcoming by resistance theorists is the overwhelming focus on overt acts of resistance while ignoring more subtle acts that do not disempower the students by shutting them out of the skills and knowledge schools do offer that can fuel social resistance and social change (Walker 1985). Thus, subordinated groups experience and resist school inequalities and domination in a variety of ways. Domination is a process that is not static, not wholly total, nor one-dimensional and neither is resistance to it. We should understand power and domination in the Foucauldian sense in that all groups can exercise power and that silence or subservience does not signify a lack of power, but a different manifestation of power (Foucault 1978). This book examines how students at a militarized charter school in Southern California interact with, make sense of, and resist militarized school structures, processes, and culture. Additionally, this book highlights how the school is a social site in which the school culture and structure intersect, as a neoliberal project, with race, class, gender, and sexualities.

Ethnographic research of schools has focused on White students (Pascoe 2007; Willis 1983) and a combination of White students and students of another racial or ethnic group (Bancroft 2009; Bettie 2003; Heath 1983; MacLeod 1987), but Latino students, in line with general US population growth of Latinos, are a growing area of school population and ethnographic study. For example, in an ethnographic study of a predominantly Latino urban high school, Flores-Gonzalez (2005) argues that high achieving Latino students do not have the burden of "acting white" (Fordham and Ogbu 1986) as they occupy different academic and social spaces at the school. Additionally, Monzo's (2005) two-year ethnography of eight Latino families examines parental choice for language instruction in California and found that choice for language of instruction was determined by structural factors such as a lack of access to information and power relations between the school and the larger community. Finally, from the same two-year ethnographic study, Monzo and Rueda (2009) argue that passing as English proficient among Latino youth is a result of racial relations in the United States. They argue that students who "pass" held an awareness of the status and power afforded to those who speak English.

Qualitative and ethnographic studies also examine the construction of gender and gender inequality in schools (see Eder 2003; Kessler et al. 1985). Thorne's (1993) work on gender and "boundary work" in schoolyard play is one of the best known examples. Thorne argues that children are not passive recipients of gender socialization, but are active in constructing and sometimes challenging gendered meanings and structures. Connell (1996) also examines how gender is constructed at schools, but specifically focuses on masculinity. He argues that masculinity is schools is a process that is multiple and complex and shaped by class and ethnicity. However, gender is not solely bounded by class and ethnicity but through other social locations such as sexuality (Butler 1995; Pascoe 2007). For example, Pascoe (2007) contends that the examination of sexuality "highlights masculinity as a process rather than a social identity associated with specific bodies" (5). Schools then are gendered and sexualized social institutions with "informal sexuality curriculum" (Trudell 1993). However, none of these ethnographies were conducted at charter schools. Educational and social processes at charter schools which are a direct outcome of neoliberal education policy, are an understudied area in the sociology of education. Thus, this book adds to the growing qualitative research on Latino and youth of color, gender, and sexualities, and examines the social processes (socialization, identity, race, gender, sexuality construction, resistance) unique to a charter school that is fully militarized.

Charter School Research

The research on charter schools can be divided into two main sectors. The first sector focuses on student enrollment and segregation (Cobb and Glass 1999; Garcia 2008; Morest 2002; Yancey 2000), while the second main sector examines the outcome of charter school policy, mainly in terms of student achievement (Berends et al. 2008; Loveless 2003; Nelson et al. 2004). As outlined earlier, charter schools exacerbate racial segregation by enrolling a disportionate amount of at-risk students of color (Nelson et al. 2000; Manno et al. 1999). Charter schools are also more likely to locate in Black and Latino communities (Henig and MacDonald 2002), and the students who attend charter schools are more likely to live in poverty (CREDO 2013). Additionally, contrary to the arguments put forth by charter school proponents, charter schools do not outperform noncharter public schools (Banchero 2010a; CREDO 2013; Loveless 2003). Thus, charter schools do little to alleviate race, ethnic, and poverty segregation and have not suceeeded in narrowing the race and class achievement gap. A missing focus of study regarding charter schools, however, is the examination of social processes such as socialization and resistance as well as race, class, gender, and sexual inequalities that students attending charter schools undergo. To that end, this book seeks to augment the literature on public charter schools, especially the literature that studies everyday school practice.

Finally, research on public education and militarization has focused on how public schools are taking on the values, goals, and characteristics of the military or the increasing connection between the US military and public education (Saltman and Gabbard 2003, 2011; Weisman 2006), but have not, to my knowledge, critically examined the daily processes, culture or resistance at a public school that is fully militarized There have been no studies that examine how race, class, gender, and sexualities are shaped by militarized schools or school settings. This book adds to the literature on public militarized schools, especially in regard to everyday school practices.

On a general level, this book examines the nexus between militarization and neoliberalism in US public education and how these macro-level social phenomena influence and shape micro-level processes at the MEI. It is a qualitative study of the relationships between neoliberalism and corporate educational reform in the form of charters and militarization.

However, this book spans across various fields of study, including neoliberalism and militarization in and of schools. This book fills the gap in not only ethnographic research in public educational settings but also the culture of a militarized public school as it examines the daily processes of students, teachers, and parents at a militarized charter school across axes of race, class, gender, and sexualities. This book is a qualitative study of a public charter school that is almost 70 percent Latino and overwhelmingly working-class. This book contributes to the literature of critical resistance theory by examining how poor and working-class cadets of color at a militarized charter school are socialized into the ranks of the military, as military enlistment is increasingly viewed as a working-class occupational choice and illuminates the processes of construction of gender and sexualities and importantly resistance.

The MEI

Current trends in US immigration and birth highlight the increasing share of the US population that is comprised of people of color and which, in the very near future, will comprise the majority of the US population. This has already occurred in California where Whites comprise only 39.4 percent of the total population (US Census Bureau 2014). These shifts in population and demographics coupled with the recent neoliberal trends in US education create an interesting social space and research arena within education in which these two phenomena merge. This is particularly true in California, the setting for this book.

MEI is a militarized charter school that is predominantly Latino. This makes the MEI an important arena of study given that Latinos are an increasing part of the US population and the military. In 2011, Latinos comprised almost 11 percent of enlisted personnel (Department of Defense 2012b) up by 4 percent from 1983 (Department of Defense 2000). The MEI is also an important area of study as it enrolls both boys and girls. This is particularly interesting considering the growing number of women in the US military,[2] the integration of women into combat positions by 2016, and the fact that women have been traditionally barred from attending military schools in the past.

What can be learned from examining a school such as the MEI? First, it is an important contribution to understanding how large social processes (such as neoliberalism and militarization) and public policies affect everyday practices of vulnerable youth and citizens. This book illuminates how militarization and neoliberalism operate within public education and their affect on socialization, militarization, and resistance of and by students. The MEI is an illustrative example of the nexus between increased militarization and the vast influence of neoliberal ideals in US society. Research on public education and militarization has focused on how public schools are militarized (Saltman and Gabbard 2011; Weisman 2006), but have not, to my knowledge, critically examined the daily processes and culture of a public school that is fully militarized, nor have any to my knowledge, examined how gender and sexualities are shaped by militarized schools or school settings or how they are resisted and transgressed.

MEI opened its doors in Fall 2003 with a grant award of $450,000 from the California Department of Education. The MEI is only the second such militarized charter school opened in California; the other school is located in Oakland, California. Retired Major Allen West founded the school. He was part of the Army Special Forces, serving in the Vietnam War, Peru, Nicaragua, and clocked many hours of active field duty. Major West has a long history with education as he has started up five different Junior Reserve Officer Training Corps (JROTC) programs across the Western United States (Roseburg and Oakland, Oregon; Tempe, Arizona, and Long Beach, California) as well as the JROTC program at a local high school. *Major*, as everyone calls him, is thin, muscular, well-tanned, and sports bleached blonde hair combed precisely into place. He has a Master's Degree in Educational Administration, has been a part of the Eastmoore School District for over 20 years, and is highly regarded and networked within the local community.

To attract students and area attention, an advertisement for the MEI was placed in several area newspapers. The day after the announcement appeared, Major West received 107 phone calls inquiring about the MEI. Initially, the MEI had only two teachers (First Sergeant Strong and Mr Jones) and enrolled 50 seventh-grade students, but expanded later that same year to accommodate 58 students.

The Eastmoore School District enrolls 7,000 students in seventh through twelfth grades in eight different schools: two high schools, a continuation school, a community day school for adult learners, an alternative school, one middle school, and two charter schools, one of which is the MEI. Only three of the schools failed to qualify for Title I funds. Title I funds are part of the Elementary and Secondary Education Act of 1965 and are intended to improve educational standards for high poverty schools and/or schools with academically struggling students (US Department of Education 1965). However, Title I funds can be reduced or redirected if schools fail to meet academic standards set by the NCLB.

Until the opening of the MEI for the 2002–2003 school year, the school district contained only one middle school. Eastmoore Middle

School enrolls over 1,300 students and because of overcrowding is on a year-round schedule. The increasing numbers of students at Eastmoore Middle School, as well as the increasing numbers of other schools in the district, was one of the reasons behind the founding of the MEI. The MEI plans to add one grade each year until fully incorporating seventh through twelfth grades. Some of the benefits to charter schools are that they place minimal responsibility on the district as each school is responsible for developing a curriculum, hiring teachers and staff, as well as other administrative duties. As a result, the MEI is a quick and easy fix to Eastmoore School District's problem of overcrowded schools and one that requires minimal time and organization on the part of the district.

The MEI is only the second militarized charter school to open in California and enrolls both boys and girls. Female students comprise 45.9 percent of cadets and 54.1 percent of cadets are male. While male cadets are the majority, it is important to note the large share of enrollment comprised of female cadets considering the underrepresentation of women in the military and the traditional barring of women into military schools and academies.

The majority of students who attend the MEI are working-class students of color. As shown in Table 1.1, the total student population at the MEI is comprised of: Latinos 66.7 percent, Whites 13.5 percent, Blacks 14.4 percent, Asian 2.1 percent, Pacific Islander 1.2 percent, Filipino 0.9 percent, and American Indian or Alaska Native 0.6 percent. Compared to overall students in the school district's schools, all racial groups except for Whites comprise a larger percentage. MEI demographics are consistent for Latinos and Blacks as both races are enrolled at school that predominately serves students of color. Approximately 29 percent of Latino students and 31 percent of Black students attended schools in 2005–2006 that were nearly all-minority[3] (Fry 2007). Only 8 percent of students at the MEI are English language learners and 92 percent of these kids speak Spanish at home. This compares to district averages of 16 and 98 percent and state averages of 25 and 85 percent, respectively (California Department of Education 2009c).

Table 1.1 Enrollment by race

	Latino	White	Black	Asian	Pacific Islander	Filipino	American Indian or Alaska Native
% of total enrollment	66.7	13.5	14.4	2.1	1.2	0.9	0.6

There are two popular perceptions of militarized schools. The first perception is that these schools are elite private institutions that feed directly into the upper echelons of the armed forces. However, students at the MEI are not funneled into elite military institutions such as West Point or the Citadel, nor does the MEI have any connection with these elite schools. The MEI is affiliated with the California Cadet Corps and the JROTC program of the California National Guard, however. The MEI graduates who choose to join the military will do so at lowest ranks completing boot camp alongside other 18-year-olds, although they may be eligible for a higher pay scale than regular enlistees. The second and most widely held perception is that military schools are a last resort for problematic students and rely upon heavy-handed discipline to set struggling kids straight. The MEI is not for students with discipline problems. Rather it promotes itself as a school for academically motivated students. Advertisements for the MEI read: "a school with above average achieving, college bound classmates." Academics is the focal marketing point of the MEI. Additionally, as it is a charter school with limited funding and space, the MEI is able to choose the students via a list of eight enrollment requirements, only two of which specifically focus on academics—a minimum 2.0 grade point average (GPA) and no failing grades in the previous school year. Not only is the MEI able to select the top applicants, but also it is also able to reject students with failing grades or discipline problems. Thus, the MEI selects the best students from the local school district by accepting only the top applicants and removing any who do not adhere to the school's standards.

The Community

The community of Eastmoore is located in Southern California and is home to the MEI. An Air Reserve Base lies just outside of the city limits and it is common to see long convoys of military vehicles traveling the freeways around Eastmoore. The local newspaper, *Eastmoore City News*, is filled with community announcements and events. Slogans such as "God Protect Our Troops" and "God Bless Our Troops" grace the front page.

The yearly Veteran Day Parade in Eastmoore is the perfect snapshot of small-town America and the daily lives of its inhabitants. The parade is filled with patriotism, flags, city council members waving to spectators from vintage cars, fire engines, police patrol cars with swirling lights, marching bands playing *Proud to Be an American*, glittery dance groups, and cheerleaders chanting "If you like it say: U.S.A.! U.! S.! A.!" More than half of the annual parade consists of the disciplined marching of local JROTC divisions, the uniformed cadets of the MEI, and the small blocks of elementary school-aged children walking hand-in-hand shouting "Left! Left! Left, Right Left!" as they make their way through the streets of Eastmoore, and the cheering parents and spectators. All of this—the parade, the local paper, and the military base—is indicative of the heavy presence and acceptance of the military, militarism, and the militarization within the community of Eastmoore. In many ways, Eastmoore, heavily characterized by the military and militaristic ideals, is indicative of numerous small towns across the United States.

Eastmoore is a solid working-class community with over 67.2 percent of its inhabitants identifying themselves as Latino, followed by 16.8 percent identifying as White, 10.1 percent Black, and 3.3 percent as Asian (US Census Bureau, American Community Survey). In recent years, Eastmoore and the surrounding communities have been caught up in a rapid growth and push east out from the larger, more expensive surrounding communities. Block shopping centers and new housing developments have sprung up along the eastern side of the freeway and have quickly filled with middle-class Whites and young families in search of affordable housing.

Although there is much new growth and development in Eastmoore, it is occurring in isolated areas mostly east of the freeway, which locals refer to as the "new side." Six concrete lanes full of commuter traffic from the "old side" separate this "new side." The "old side" displays the scars of working-class or poor neighborhoods: grassless yards, chain link fences, broken down automobiles, barred windows, empty lots, boarded up store fronts, and pothole-filled roads. It is the "old side" that houses the MEI and the majority of its cadets.

Data and Methods

I initially learned of the MEI from a feature story in a local newspaper. I had just moved back to the United States and was active in the antiwar movement both stateside and abroad. I was shocked to learn that 11- and 12-year-olds were attending a militarized school that was publicly funded. I wanted to know more. Having no network or contacts within the Eastmoore School District, I placed a phone call directly to the principal and commandant, Major West. I explained that I was a sociologist from a nearby university, very much interested in alternative forms of education, and that the MEI especially intrigued me. Although I was not upfront with Major West about my antiwar stance (we actually never discussed the war in Iraq or Afghanistan), I was open about my curiosity of a militarized school, how it was structured, the social life of the school, and why students and parents would choose to attend the MEI. Major West was very candid and open. He was extremely receptive to my proposal to study the MEI, including observing and interviewing both staff and students. Major West invited me to meet him in his office the following week.

Major West's office was simple. A large US flag was stretched across the wall behind his desk and faded pictures from his tour in Vietnam adorned the walls. Major West has been working with JROTC programs up and down the West Coast for over 20 years. Major West explained his reasons for opening the school, his hopes and goals for the school in the future. He also showed me a local television news clip about the school. The initial part of my meeting with Major West felt very rehearsed, much like a sales pitch given to prospective students, curious parents, or community leaders. Eventually, the conversation turned to the outdoor and overnight fieldtrips that are a vital part of the MEI culture. Coming from a White, rural, working-class background, I was able to knowledgeably discuss camping, hiking, and gun culture with Major West. It was at this point that there was a noticeable drop in defenses by Major West and an ease that remained for the rest of the meeting. I believe that my White, working-class background and my ability to discuss gun culture and outdoor life with Major West aided immensely in my gaining access to the MEI.

My research is based on multiple sources of data and methods, including initial observation, three years of participant observation, informal conversations in a variety of settings, and formal interviews. I began my research at the school in spring of that year, and I stayed in close contact with the school for four years. I am still in contact with many of the MEI students and parents. I stop by the school from time to time, meet up with students and parents at restaurants and cafes, and regularly receive and send greetings and updates through online social networking sites.

In the first few months at the MEI, my research consisted of simple observation, as I wanted to learn as much as possible about the school, its culture, students, and staff. After several initial months of observing as much of school life as possible (classes, morning formation, marching drills, flag ceremonies, cannon ceremonies, and day fieldtrips), I conducted my first focus group with eight seventh graders. While this focus group did yield interesting data that brought to light several important themes, it was rather chaotic, as might be expected from a room of boisterous seventh graders days before summer break. After this point, I rarely conducted group or paired interviews. Additionally, the students who participated in this initial focus group were chosen by Major West and were not reflective of the school's demographics. In these beginning stages of my research, I was not fully immersed at the MEI, and I believe this lack of full immersion is why Major West chose the students as these students were all outgoing, cheerful, and positive about their experiences at the MEI. I did not gain total immersion until I began my participant observation in the second year, in which I helped the school to set up a website, and the third year in which I served as an office aide, lunchtime monitor, and chaperone.

As a qualitative research project, this study utilizes the grounded theory approach as proposed by Glaser and Strauss (1967), relying on inductive reasoning, comparison, and a general focus on categories and codes to build a theoretical explanation of social life at the MEI. The bulk of my research consists of participating in the school's daily life and getting to know the students, staff, and parents and understand how they organize and interpret the social world at the MEI. Having a sense that students would view me as an authority figure and thus be less likely to open up to me, I purposely did not take notes while at the MEI. At the end of each day, I recorded my field notes, focusing on how students and staff negotiated, regulated, and resisted notions of gender, race, sexualities and militarization. There were several times that I sat in my car in the school parking lot, hurriedly scribbling down particularly meaningful observations. I did not include any theoretical ideas in the field notes in order to keep the field notes a true depiction and representation of life at the MEI. Abstract thoughts and theories were kept in a separate file. As my field notes and time at the MEI accumulated, themes began to emerge from the data. Field notes were organized and indexed into emergent themes from my observations and I eventually found myself recording similar observations and notes again and again.

I was heavily involved with the MEI during my third year of research. I served as a lunchtime monitor, chaperon at school functions and fieldtrips, and completed small tasks for staff and faculty such as data entry, proctoring exams, and photocopying. It was also during the third school year that Major West retired and Commandant[4] Wilson took over. There were a lot of changes in the third year, but this was also the time I felt the closest to students, parents, and staff. It was during the third year that I truly felt that I was part of the school as I shared gossip with the school secretary and aide, blew up balloons, ate doughnuts with parents, listened to students' worries, and offered advice in the cluttered supply room.

I logged close to 50 hours of observation and over 600 hours of participant observation at the MEI.[5] Both observation and participant observation occurred in a variety of settings, including daily school life, fieldtrips, military drill competitions, sporting events, and after school events such as Parent Day and the end-of-the-year Military Ball. I interacted with students before and after school, on fieldtrips, and during their free time between classes and at lunch. I became very close to the 12 students who were part of the military drill team as I was one of two volunteer chaperones at a weekend-long military drill state finals. The students in this group were boisterous, friendly, and took military drill very seriously. The drill team consisted of eight girls and four boys and was racially diverse.[6] In addition to the military drill team, I also became quite close to a small group of cadets, 5 of whom were Latino/a and 1 Asian. I got to know these students through our lunchtime interactions and the group's general curiosity toward me. These students were not highly integrated into the military culture of the MEI; none held any formal student positions or high military rank. They received decent grades, mercilessly teased each other, and kept themselves out of trouble and the commandant's office. During my three years at the MEI, I also become close to three other students who often confided in me and asked for my advice: a White male student, a Latino, and Latina.

Additionally, I held reoccurring conversations with all the teachers as I met them in the hallways, the large storage closet, teacher's lounge, or other school activities and locations. I regularly chatted with administrators[7] (school secretary, security officers) and the commandants throughout my time at the MEI. This usually occurred in the main office as I completed small tasks for the school secretary or a teacher. During my time at the MEI, the teaching staff changed every year as teachers moved on to better-paying jobs both inside and outside of teaching. For the most part, the teaching staff at the MEI was racially diverse.[8]

In addition to interacting with parents at military drill competitions, fieldtrips, and other school events, I became very close to a small yet tight-knit group of parents[9] that regularly volunteered at the school. This group of parents coordinated fundraising events for the school, such as a recycling drive and sold hot and cold lunches twice a week as the school lacked food services at that time. I regularly helped with these events and took my turn bringing morning pastries for the group. I often spent the beginning of each day at the MEI chatting and gossiping with parents and administrative staff over these pastries and coffee. The majority of the parent volunteers were female, White and working-class. I am not sure why the majority of parent volunteers are White, but it could be that White parents had more agency at their jobs and could organize their schedules in order to volunteer at the MEI. However, several of the parents were unemployed and volunteered at the school in hopes of securing an aide position in the following school year. Black and Asian parents were markedly absent from the group of parents that volunteered. Since I rarely interacted with Black and Asian parents, I cannot assume the reason behind their absence from the volunteer work at the MEI.

As ideas and theories emerged from the data, I returned to the school, the students, staff, and parents for validation of my observations and clarification of questions. I did this through additional participant observation, but also through conversations with students, parents, teachers, and staff, as well as formal, semistructured interviews. Formal interviews were conducted, for the most part, at the school, but I did on several occasions meet staff or parents off-campus, at home or at a local coffee shop. Formal interviews were on a one-on-one basis, and only twice did I allow students to be interviewed in pairs or groups. In total, I conducted 16 one-on-one formal interviews and two paired or group formal interviews with students over the course of my research. I chose students who were heavily involved in the school and military culture, as well as students who did not particularly enjoy the MEI or military culture. I also formally interviewed students who approached me and asked to be interviewed. I formally interviewed 14 boys (six of these were part of the initial focus group and another six came from a group interview during lunch) and 9 girls. Apart from the two female students who participated in the focus group, the girls were all interviewed on a one-to-one basis. In regard to race, the formal student interviews consisted of 14 Latino students, 8 White students, 4 Black students, and 1 Asian. The racial composition of formal student interviews closely resembles the student population at the MEI although Black students were slightly underrepresented in my formal interviews as I found them the least responsive to my attempts to talk to them and get to know them. This could be due to the fact that they viewed me as an outsider due to my racial status.

In addition to formal student interviews, I interviewed parents, teachers, and administrators. I interviewed five of the seven teachers.[10] I also interviewed the founder of the MEI and commandant for the first two years, Major West, and his successor, Commandant Wilson. Finally, the parents who I interviewed were all women who were highly involved with the school and often volunteered at the school in varying capacities: as chaperones, at fundraising events, and military drill competitions. Two of the parents were highly involved in the Parent Teacher Association.[11] Parents were the most difficult to formally interview. As the MEI serves a predominately working-class community, most parents of cadets at the MEI worked long hours and on weekends. The hectic schedules of full-time work and family responsibilities made scheduling time to talk difficult, if not impossible. Manytimes, interviews were scheduled and then canceled at the last minute. Although the number of parent interviews is small, it is important to remember that formal interviews were utilized as a way to clarify emerging themes and trends. The bulk of my information and interaction with parents came from my daily involvement at the school and interaction with about 25 parents at the school, events, and fieldtrips and the close interactions with the six parent volunteers.

All interviews were semistructured, lasted from 45 minutes to an hour, and were taperecorded. The student, teacher, and staff interviews were conducted during school hours in various venues around the school (empty classrooms, the teachers' lounge, and the large and jumbled storage closet). Parent interviews were conducted off campus at a location chosen by the parent, usually their home or at a local coffee shop or restaurant. I interviewed parents who were heavily involved in the school, but also parents of students who I became close with over the course of my research. Overall, I interacted in various settings and contexts with all the teachers and administrative staff at the MEI, close to 25 parents and hundreds of students of varying grade levels, races, and both sexes.

Organization of this Book

Chapter 2 examines the relationship between militarization and public education. I argue that militarization, particularly within public schools, is strengthened through neoliberal policies and are most frequently applied to public schools located in poor or working-class neighborhoods and communities of color. As these schools are often labeled as failing by school districts and states, neoliberal and militarized policies such as JROTC, the GI Bill, Troops to Teachers, the Federal Development, Relief and Education of Alien Minors Act, Department of Defence STARBASE, STARBASE 2.0 and policies are presented as disciplinary solutions to fix failing schools. However, as the US military is dependent on volunteer enlistment in order to ensure the implementation of global neoliberal practices, such militarized educational reforms and programs are nothing more than recruitment strategies that exploit those youth marginalized by the very neoliberal practices they in turn defend upon enlistment.

Chapter 3 examines the reasons parents and students enroll in the MEI. I argue that the MEI and the Eastmoore School District successfully took advantage, through neoliberal militarized logic, the structural gaps and inequalities in the local school district, namely underfunded, overcrowded, and violent schools, to open a militarized charter school. The racial tension, presence of drug and gang activity, and academically failing and overcrowded local schools push parents to look for educational alternatives. This coincides with parents' and teachers' perception that the militarized discipline-based structure of the MEI will keep their children safe from violence and ensure academically successful futures. Additionally, the militarized uniforms confer social status within the working-class Eastmoore community, marking MEI cadets as different and superior. The fact that the alternative to failing local schools is a militarized charter school was not the primary factor in the cadets' and parents' decision to attend the MEI, but rather a consequence, I argue, of social inequality. I contend that parents "buy in" to the logic of neoliberal choice and discipline as it offers agency and an alternative to the inequitable and violent school district. Parents consent, in Gramscian terms, to militarized education and choose to enroll their children in a militarized charter school. However, this is more indicative of an underfunded and inequitable education system that limits the choices and participation of marginalized groups rather than true empowerment and agency.

Chapter 4 illustrates how military recruitment practices focus on poor and working-class youth of color and how militarized policies, programs, and schools such as the MEI serve to promote the recruitment agenda of the US military. Specifically, the discursive use of militarized symbols (the school mascot, the cannon and flag ceremony, the "Rough Ride Ration," and the uniform) and language (school motto, hierarchical structure of the school, school fieldtrips) at the MEI militarizes the youth and shapes their identities and subjective understanding of the social world. Thus, as I argue, cadets view the military as an attractive and equally beneficial career choice as higher education and this increases the chances that MEI students will be open to military recruitment. However, people of color overwhelmingly occupy the lowest ranks of the US military and receive the least amount of training that is transferable to the civilian occupational market. Thus, while poor and working-class youth of color come to view the military as an opportunity and a wise career choice, it is, in fact, not.

Chapter 5 discusses how masculinity is constructed at the MEI for both boys and girls. I argue that hegemonic militarized masculinity is exemplified at the MEI through the condonement of violence vis-à-vis a warrior hero archetype. However, the construction of hegemonic masculinity at the MEI is bounded by race and gender as only White males are allowed to capitalize on this particular form of masculinity. Girls and Black boys who attempt to access hegemonic masculinity are particularly severely sanctioned. Not all cadets want to enact hegemonic masculinity as illustrated by normatively feminine girls at the school that actively resist militarized masculinity.

Chapter 6 argues that militarization of the MEI enforces normative gendered and sexualized processes and behaviors such as sexual scripts for flirting and dating as well as homophobia and homophobic slurs. While the MEI is a militarized institution of heteronormative and gendered practices, it is also a queerly resisted and contested space (through heterotopias, carnivalesque moments, and subaltern counterpublics) in which students actively challenge these normative standards and create their own sexual rituals, practices, identities, and spaces. This chapter extends the explanatory potential of resistance theory by incorporating aspects of queer theory.

Finally, Chapter 7 revisits the themes and topics discussed in the previous chapters and lays out the theoretical importance of this book. In this chapter I argue that this book is an important contribution to understanding the nexus between neoliberal educational policies and militarization by highlighting how these larger social processes affect the daily lives of working-class youth of color who attend not only MEI but similar militarized schools and programs. I argue that social structures such as a lack of quality and equitable education, as well as inequalities of race, class, gender, and sexual identity create social space for the emergence and growth of militarized public education. But what must not be forgotten is that students who attend the MEI and other militarized schools are not passive dupes, but agents who resist heteronormative, gendernormative, raced, and classed structures of militarization and neoliberalism. To close the chapter, and taking note from the students at the MEI that actively challenge and resist militarization of their identities, education, and everyday lives, avenues for activism and resistance of militarization and neoliberalism are outlined. Drawing on Kirk's (2008) work, I outline strategies for resisting militarization across personal, community, institutional, and global levels with specific strategies to combat militarization of US public schools.

2

Schools in the Crosshairs: Neoliberalism, Militarization, and Public Education

It is a sunny fall morning in Southern California. US flags rustle from light posts and the streets of Eastmoore are lined with lawn chairs, rolling coolers, and people jostling for sidewalk space. It is the annual Veteran Day Parade in Eastmoore, and amid the local politicians, fireman, squad cars, and equestrian units are hundreds and hundreds of children. Some of these children form marching bands sputtering out slightly-out-of-tune patriotic songs such as "My Country 'tis of Thee" and cheer squads that encourage the crowd to yell in unison "U.S.A.! U.S.A.!," but most of the children are clad in military uniforms and formed into neat lines marching together as human blocks through the city. Wave after wave of khaki and dark blue, line after line of marching youth: Junior Reserve Officer Training Corps units, California Cadet Corps brigades, private military academies, the color guard from the MEI, and even a vulnerable looking group of fourth and fifth graders holding hands shouting in singsong unison "Left! Left! Left, Right, Left!"

The United States is a warrior state. Not a state at war but a state *of* war. Since 1776, the United States has engaged in war or significant military action for 80 percent of its history and has not had one year since 1941 when the nation was not involved in some type of significant military engagement (Brandon 2003, 1820). The United States spent $682.5 billion dollars on military expenditures in 2012, an astounding 39 percent of total world military expenditure for that year. In 2012, the United States spent more than the next ten countries combined and a little more than four times that of China ($166.0 billion), the country with the second largest military expenditure (SIPRI 2013b). The total number of US military personnel in 2012 was over 3.6 million, making the military the largest employer in the United States (Department of Defense (DoD) 2012b), and the military owns or leases 5,211 geographic sites domestically and globally (DoD 2012a). The United States was the largest arms exporter in all years from 1990 to 2012 (except for two years 2001–2002) (SIPRI 2013a) and controlled 30 percent or more of the market from 2003 to 2012 (Holton et al. 2013). The United States is undoubtedly a nation of war.

In a militarized nation such as the United States, the perpetual state of war and war readiness is normalized as military strength and violence are equated with diplomacy and national security. In the United States, military experience carves men out of boys, educates its citizens, and fuels the economy. The United States views the world through the crosshairs of a military sight and, as C. Wright Mills warned over 50 years ago, has "a military definition of reality" (1956, 191) in which the values of the military are understood as not only ordinary but also right and just.

Militarization, or the alignment of institutional, cultural, linguistic, governmental, and economic forms with militaristic values and beliefs, is the "step-by-step process by which a person or a thing gradually comes to be controlled by the military or comes to depend for its well-being on militaristic ideas" (Enloe 2000, 3). However, the more the society is in sync with military values, the more accepted and ordinary militarization becomes. Thus, institutions, people, and cultural practices are legitimized through association with the military. For example, patriotism is defined as supporting the troops as opposed to critical dialogue or democracy. Bravery and strength are linked with martial violence instead of diplomacy. Military discipline is the solution for inner city violence, troubled youth, and failing schools, and war is a form of entertainment (Mann 1992) as video games, children's toys, and films glorify military history, values, and practices. In such societies, as was described in the scene of the Eastmoore Veteran Day Parade, it is not unusual to see children playing at war dressed in the drag of military units marching through the streets of small towns.

Militarization is a raced, classed, and heterogendered project. For example, domestic militarization is found in the violence used on and against people of color (Gilmore 2000) and also "seen in the rise of the prison-industrial complex, the passing of retrograde legislation that targets immigrants, the appearance of gated communities, the widespread use of racial profiling by the police, and the ongoing attacks on the welfare state" (Giroux 2003, 39). The acceptance of militarization is also situational and bound by social context. The militarization of children and childhood is normalized within the United States (think children summer camps at military bases, Junior Reserve Officer Training Corps

(JROTC) programs, and the Boy and Girl Scout programs), but criticized heavily when it occurs abroad especially in non-Western countries such as Somalia or Uganda (Macmillan 2011). The organized violence of military service by young people is respected and valued and constructed in direct opposition to the disorganized violence of youth gangs—viewed as dangerous especially if these gangs are comprised of poor and working-class youth of color (Basham 2011). Even gender is not beyond the reach of militarization as "militarism and militarization redefine both masculinism and femininity, alongside a hyper-sexuality and neoracism that construct new-old racialized gender formations" (Eisenstein 2007, 18).

This chapter examines how militarization and neoliberalism intersect to shape US public education specifically focusing on public K-12 levels. The central theme of this chapter is how the relationship between militarization and public education is strengthened by neoliberal political and cultural ideologies. I argue that neoliberal policies and militarization are forces that work in tandem and are most frequently and vehemently implemented within public schools located in poor and/or working-class neighborhoods and communities of color. This chapter begins with a brief overview of the development of the US public education system outlining the initial formation of public education as a public good with the purpose to build consensus and a stronger nation through national identity. This is followed by an in-depth examination of the relationship between neoliberalism and militarization and the effects on public education in the United States. I argue that militarization of the public education system is a strategy for military recruitment that specifically targets those most vulnerable to neoliberalism: poor and working-class youth of color.

Development of the US Education System

Historically, there have been three broad goals in the US education system: democratic equality, social efficiency, and social mobility (Larabee 1997). Each of these goals arose in specific historical moments during the development of the US education system. The first goal, democratic equality, prepares White youth for citizenship and was the primary goal for the founding of the common school in the mid-nineteenth century (Larabee 1997, 43). In order to accomplish this, not only did schools have to be standardized to some extent, but they also needed to provide equal access (to provide enough schools) and equal treatment in the hopes of assimilating a growing population of immigrants. At this time, education was viewed as a public good and integral to the building of a strong nation.

The second goal of US education system is social efficiency, or the training of workers. At the end of the nineteenth century and beginning of the twentieth century, the school curriculum shifted from a focus on knowledge to a focus on skills and training and became more responsive to the occupational structure of the economy (Larabee 1997, 47–48). Schools became increasingly stratified reflecting the job market and differentiated levels of skills and training needs. Although vocational-based training readied students for the job market, it was still viewed as a common good, as an educated and trained citizenry proves beneficial to the strength and well-being of the nation as a whole (Larabee 1997, 49).

The third and final goal, social mobility, rose in conjunction with social efficiency (Larabee 1997, 58). Education became more differentiated, and educational attainment began to translate into job market opportunities. Education was viewed as a commodity consumed by students that provided them with competitive advantages in the society. Thus, the level of education or credentials earned became more important than the acquisition of knowledge, and individuals pursued skills or credentials that were associated with a higher social status as a means of social mobility commonly referred to as "contest mobility" (Turner 1960). Education became a private good, a zero-sum game where some students excelled at the expense of others, and increased stratification and competition made the first historical turn toward the logic of the market (Larabee 1997, 50–52).

In the 1940s, educational leaders began to question the inequitable distribution of educational resources. In 1954, the Supreme Court ruled that racially segregated schools were inherently unequal. Desegregation in US education system began shortly thereafter in an attempt to equalize school opportunity. In 1966, the *Equality of Education Opportunity* report commissioned by the US Department of Health, Education, and Welfare was released and highlighted the inequality in the US education system along racial lines. Additionally, cities and states released their own reports of educational inequality. All of these reports came to the same conclusion that poor, students of color were falling further behind their econimcally advantaged and White counterparts in academic achievement with each passing year.

Thus, the American dream of social mobility embraces the ideas of individual free choice and unbounded possibility that are essential to the model for capitalistic societies like the United States, and not coincidentally in line with neoliberal theoretical values that were also beginning to take hold in the 1970s. However, social mobility is only possible if some students fail. The US education system is a contradiction and has been since its inception. Espousing equality of opportunity and the ability for all to "make it" if they work hard enough, the system cheats players by constructing more roadblocks for some than for others. Obstacles structured around gender, race, sexualities, abilities, and class such as unequal school funding, segregation (Kozol 1991, 2005; Leonardo and Grubb 2014), bias in testing (Goslin 1967; Leonardo and Grubb 2014; Taylor 1991), and tracking (Leonardo and Grubb 2014; Lucas 1999) ensure that particular students are unsuccessful. Thus, the "one best system" founded by early education leaders in the nineteenth century had failed.

In 1981, Ronald Reagan's Secretary of Education, T. H. Bell, created the National Commission on Excellence in Education with the sole purpose of examining the quality of education in the United States. The resulting report, entitled *A Nation at Risk*, argued that "the educational foundations of our society are presently being eroded by a rising tide of mediocrity that threatens our very future as a Nation and a people" (US Department of Education 1983, 1). The report also stated that test scores and educational standards were falling and US students were falling further and further behind students in other countries. It recommended raising education and graduation standards as a way to reverse the trend of mediocrity. Emerging with this perceived failure were new theories of the one best system explicitly based on neoliberal theories and values. As discussed in chapter 1, moments of crisis become moments of opportunity to implement neoliberal reforms referred to as "disaster capitalism" (Klein 2007), "accumulation by dispossession" (Harvey 2005), or "smash and grab" capitalism (Saltman 2007). This was one such moment.

Regardless of the specific name for such predatory implementation of neoliberalism, the war on the US public education system had commenced. Under neoliberalism, education is now defined as a private consumable good dictated by choice and individualism, and failure is defined in terms of individual shortcomings rather than structural inequalities, while militarization steps in to discipline those students who slip through the neoliberal competitive cracks. The increasing militarization in public schools and programs is the focus of the next section in which I argue that these schools and programs serve as educational camouflage for recruitment strategies that target poor youth of color.

What's the Connection? Neoliberalism, Education, Militarization

Although the United States has a long history as a militarized society, US militarization has increased dramatically with the rising influence of neoliberalism. Since the 1970s, neoliberalism has become the central economic and political organizing principle across the globe. As outlined in chapter 1, neoliberalism is based on capitalistic, free-market policies and espouses competition, accountability, choice, efficiency, and discipline. Almost all states have implemented some form of neoliberal theory or practice whether voluntary or under coercion (Harvey 2005). While neoliberalism is, at its heart, an economic theory, it is also a cultural project (Giroux 2004) espousing individual freedom and rights. However, such individual rights are only guaranteed through the freedom of the market and only enjoyed by those who have the income or security to do so. Freedom, as the base fundamental idea of neoliberalism, is compelling and powerful and is termed a "cultural pedagogy of neoliberalism" (Giroux 2004) that teaches the fundamental values of the market. "As a seductive mode of public pedagogy, neoliberalism extends and disseminates the logic of the market economy throughout society, shaping not only social relations, institutions, and policies but also desires, values, and identities in the interest of constructing 'the citizen-subject of a neoliberal order'" (Brown 2005, 151–152).

What is more, neoliberalism defines or conflates democracy with economic freedom—the freedom of the individual to participate in the marketplace. According to neoliberal discourse, democracy, or rule by the majority, threatens individual, market-based freedom. Thus, neoliberalism is at its very core antidemocratic. To guard against rule by the majority, "the neoliberals have to put strong limits on democratic governance, relying instead upon undemocratic and unaccountable institutions (such as the Federal Reserve of the IMF) to make key decisions" (Harvey 2005, 69). Thus, there is an inherent contradiction within neoliberalism. On the one hand, the state should put institutions and practices in place to ensure free enterprise, but contradictorily, the neoliberal state should not use its power beyond this scope to limit such things as democratic governance. What is more, as the state is only concerned with ensuring a welcoming business climate, neoliberalism makes individuals solely responsible for personal failings due to structural inequalities, many that stem from neoliberalism in the first place. Militarization is one outcome of neoliberal practices and policies as people are pushed to enlist in the military as a means of economic survival and stability.

Additionally, to ensure a favorable business culture, the neoliberal state utilizes its power to suppress dissent or opposition to corporate power such as coercive legislation, policing, and state surveillance. Thus, the corporate state becomes the punishing state, but it is also a racialized and classed punishing state, one that increasingly criminalizes social tensions, social problems, and those citizens who do not benefit the neoliberal and corporate state. Accordingly, any social inequality is the result of individual choices, not structural inequalities. The focus of neoliberalism on radical individualism depoliticizes social inequalities, such as race, glass, gender, and sexualities, and boils them down to personal choice in the marketplace. This is what Giroux (2009) calls the "biopolitics of neoliberalism and disposability" or "an education, cultural, and political discourse that has laid the groundwork for a set of practices and policies in which young people are increasingly defined through market-driven ideas, social relations, and values that are predatory in nature and punishing in their consequences" (xiii). Thus, large portions of the population are viewed as of no use to corporate expansion and thus disposable, regulated to objectionable working and living conditions, enrolled in underfunded and neglected public schools, and sent to the frontlines. As Giroux suggests, youth "have come to be seen as a source of trouble rather than a resource for investing in the future, and in the case of poor black and Hispanic youth are increasingly treated as either a disposable population, cannon fodder for barbaric wars aboard, or the source of most of society's problems" (2009, 18).

Surveillance and militarization are the ways in which neoliberalism aims to control populations of disabled citizens who threaten neoliberal "democracy" and "freedom(s)." In education, this means increasing security measures in schools such as on-campus police officers, security technology, and zero-tolerance policies. It also means increasing discipline of unwanted populations through JROTC programs and other militarized school policies and structures (detailed later in the chapter).

Neoliberalism is not solely a domestic project, it is also a global project witnessed through the attack on the welfare state, austerity measures, privatization schemes, and policies implemented across the globe through the IMF and World Bank. In particular, the IMF, under the auspices of development, conducted variable-rate loans to poor countries that mire them in insurmountable debt. These loans carried with them conditions on these poor countries in the name of increasing efficiency in the global economy and reducing state expenditure in order to repay debt, as well as reducing barriers to trade and increasing privatization of state-run utilities and programs such as healthcare and education (McMichael 2008, 128–144).

These "rollback" processes coincided with "rollout" processes where the state utilizes coercion and punitive discipline through the police and the military in order to enforce these new state policies (Cabezas et al. 2007, 6–7). Thus, neoliberalism is not alone in its ascent to primary corporate globalism and goes hand in hand with militarization (Enloe 2007; Saltman 2003). Global markets are literally forced to open as neoliberal privatization policies are backed by national force, while at home more and more social institutions and practices come to be militarized by taking on the values and goals of the military. What is more, as neoliberalism is more and more dependent on the military to enforce its goals and reforms, as well as control the masses that threaten neoliberal freedom through majority rule, the military is becoming another institution for corporate greed. Private corporations have penetrated the military and have taken on work (for profit) that soldiers used to do (such as preparing meals or doing laundry).

Additionally, private mercenaries or corporate warriors (Singer 2011) are hired to fight wars and are the second largest presence in the forces in Iraq (Eisenstein 2007, 61), making profits the new patriotic motivation for warfare. Neoliberalism needs the military and militarization, in turn the military is corporatized and another arena for profit making. To keep all this churning, not only must society and its institutions be militarized, but society and the threat of democracy must also be kept under the thumb of neoliberalism. All of this relies on a fully funded and staffed military force.

In 1973, the United States ended conscription and turned the military into an all-volunteer force reliant upon recruitment to keep the ranks filled. The move from the draft to the volunteer force is credited not only to social discontent and an unpopular war but also to a group of neoliberal economists such as Milton Friedman and Walter Oi who greatly influenced Nixon's presidential campaign and his rationale for ending the draft (Bailey 2007). Accordingly, the abolishment of the draft, it was argued, would alleviate concerns of unfairness (lottery draft, service exemptions), and individuals would join the military based on rational, economic interest. The military would compete on the job market for recruits through benefits and pay much like other employment opportunities (Moskos 2005) and rely on recruitment strategies to maintain enlistment.

The Pentagon set aside $913.8 million for recruitment in 2013 (Emmanuel 2012) and utilizes different basic strategies for recruitment premised upon contacting and having access to youth. For example, the Armed Services Vocational Aptitude Battery is sold to students as an aptitude or career placement test, but the results of the test, along with the test takers' private information, are forwarded to the military. Military recruiters use this information to identify recruitment leads who meet the basic requirements for military service. Additionally, the Youth Attitudinal Tracking Survey was used from 1975 to 1999 to examine enlistment propensity, advertising awareness, recruiter contact, and slogan recognition of youth. The military also relies on behavioral and psychological research to identify factors that encourage teenagers to enlist

(Eighmey 2006). In addition to the examples outlined above, the military employs legal access, advertising, and militarization of schools and educational programs to contact and recruit youth for military service.

First, the Armed Forces make use of the legal system to gain and ensure continued access to youth. For example, the 1996 Solomon Amendment blocks federal funding to universities that do not allow military recruiters or Reserve Officers' Training Corps (ROTC) on their campuses. Another example began in 1999 when the Secretary of the Army, Louis Caldera, developed and implemented the Hispanic Access Initiative. This plan expands Army ROTC to university campuses with high Latina/o enrollment and pays "instructor salaries, scholarships, marketing dollars and operating expenses at each of these schools where traditionally 'underserved' [in military recruitment and enlistment] populations enroll in college" (Boje 2003, 4).

Additionally, in 2002, the US government further pushed its recruitment agenda of youth through legal means. The NCLB Act went into affect and blew open the doors of public high schools by granting military recruiters full access to the personal information of high school juniors and seniors, which included their names, addresses, and phone listings (Paige and Rumsfeld 2002). Like the Solomon Act, schools that deny recruiters' access to students or campuses risk losing all federal financial aid. It is not without notice that it is urban schools that predominantly serve students of color that are most harmed by a lack of financial funding. Such legislation comes at a time when most schools are scrambling for resources and pushing up against bulging class sizes. Additionally, poverty and joblessness that many poor and youth of color experience, combined with the promise of money for college, job skills, and job training made by recruitment officers, make them easy targets for military enlistment.

Advertisements are the second basic tactic the Pentagon utilizes for recruitment and is essential to sustain the volunteer force. This is not a wholly new phenomenon as the US military has a long history of using advertising to boost recruitment.[1] The US military utilizes four different advertising agencies to market the military toward specific racial and ethnic categories (Venetis et al. 2011). For example, in 2000, in order to increase the recruitment rates of Latina/o and Black populations, the army subcontracted with two public relations firms specializing in marketing products to these two populations (Schaeffer-Duffy 2003). This resulted in the US army's "Take it to the Streets Campaign" focused on Black Americans, a recruiting program that uses hip-hop culture as a vehicle for military recruitment (Chery 2003). The US military also uses advertising to increasingly recruit women, although it, until very recently, refrained from depicting women in combat roles (Brown 2012). Additionally, the US army partnered with NASCAR, spending approximately $15.4 million in 2011 (Lehman 2011). In 2013, the US army spent just under $9 million to produce and air reality-style commercials that were 30 minutes long and followed the lives of 10 young women and men as they contemplated enlistment. These commercials were aired in areas where the army has struggled to attract enlistees, such as New York, Chicago, and Washington (Elliott 2013). All four branches of the military also have official channels on YouTube.com where a variety of high-paced video clips, flashy advertisements, and gritty, raw footage can be viewed by interested or curious youth. While advertising and legal access to youth are important recruitment strategies, public schools are one of the prime areas to contact and recruit youth for military service. Public schools hold large populations of youth for extended periods of time, which ensures the military a captive pool of possible recruits. In the following section, militarization of public education and programs is examined. I argue that these programs and policies serve as targeted recruitment strategies of vulnerable, disadvantaged youth.

Examples from Frontlines

The Junior Reserve Officer Training Corps

The JROTC is the first historical link between the military and public education. The JROTC program began under the passage of the National Defense Act of 1916 and was initially an army program with the primary purpose of "dissemination of military knowledge and values among the U.S. secondary-school population" (Taylor Jr. 1999, 41). The ROTC Vitalization Act of 1964 expanded JROTC to all branches of the military and allowed retired officers to be instructors, and a 1972 court ruling opened JROTC to girls (Taylor Jr. 1999, 43). One of the biggest surges in JROTC numbers came, interestingly, after the 1992 Los Angeles riots when President George H. W. Bush and the chairman of the Joint Chiefs of Staff, General Colin Powell, challenged the JROTC program to double its numbers in five years with a new initiative aimed at troubled youth (Taylor Jr. 1999, 44). This initiative came at the time when concern over violence in schools and dangerous poor youth of color was at a fevered pitch and coincided with the implementation of another militarized educational policy—zero-tolerance (discussed later in the chapter). Utilizing such stereotypical and racist discourse frames poor youth of color as dangerous, unruly, and in need of discipline and control (Lipman 2011a).

Thus, JROTC programs become a solution to underfunded schools filled with students of color in need of discipline and the military to "set them straight." It is interesting that the solution to violence in schools is a program funded by one of the most violent organizations—the Pentagon. This may seem as a contradiction of sorts, but is actually in line with militarized understanding of violence, race, and class. The militarized state wields institutionalized and legitimate violence through discipline while poor youth of color who react to racist and classist social structures in aggressive ways are thought of as dangerous.

In 2006, the Pentagon set the goal of expanding the program to 3,570 units by 2014 and upped the number to 3,700 in 2009 with the National Defense Authorization Act. It was also in 2006 that the Pentagon pinpointed poor youth of color for the JROTC program. The Pentagon set a goal of 20 percent of expansion of the JROTC program to be in "educationally and economically deprived areas" that serve "at-risk" youth (Expect More 2006). JROTC programs are more frequently found in low-income communities of color (Berlowitz and Long 2003; Lipman 2004). As of 2013, there were 3,402 JROTC units nationwide with over half a million cadets (Jones 2013). The JROTC also operate abroad with additional units in Japan, Korea, Germany, Italy, Belgium, the Netherlands, Great Britain, America Samoa, the Northern Marianas, and the Panama Canal Zone.

A rising trend is not just schools that host JROTC programs, but entire schools organized as JROTC units. These public military academies are partially funded by the DoD, and all students who attend these academies are enrolled in the JROTC program. Additionally, military personnel work at the schools, and military courses such as military history are required curriculum (McDuffe 2008). The first public military academy opened in 1980 in Richmond, Virginia. The founding of the Franklin Military Academy was argued to help "improve discipline and general attitudes of students" (Franklin Military Academy website). The Chicago public school district has the most military academies, with six schools representing all four branches of the armed forces (Banchero and Sandovi 2007). There are currently 21 military academies nationwide, with more opening in the 2014 and 2015 school years (militaryschoolusa.com).

Politicians and policy makers frame the JROTC as a discipline-based resource for at-risk youth that builds leadership and citizenship skills rather than a recruitment tool for the military. However, JROTC is funded through the recruitment budget of the Pentagon (McDuffee 2008); students who partipate in JROTC programs are more likely to enlist (Biggs 2010) and studies show that 50 percent of JROTC cadets join the military after high school (Lutz and Bartlett 1995; McDuffe 2008). In 2000, former

Defense Secretary, William Cohen, testified to the House Armed Services Committee that the JROTC is "one of the best recruiting devices that we could have" (as quoted in Jones 2013).

The Defense Department guidelines for the JROTC specifically state to target "at-risk" students in poor, urban schools (Berlowitz and Long 2003; Galaviz et al. 2011). The targeting of students of color through JROTC cannot happen without the support of local school districts. JROTC is sold to underfunded urban schools as a way to subsidize and expand their sparse educational programs. Contracts between the DoD and the school districts hosting JROTC programs include partial salary payment for JROTC instructors, support for uniforms and equipment, reimbursement for administrative and transportation expenses, and curriculum development (Taylor 1999, 6). However, the schools must pay the remaining salary for the instructors and provide adequate classroom and drill space (Taylor 1999, 6). While JROTC promises educational opportunities (through the possibility of earning college elective credits) and adventure, the cost of JROTC may be too high for already struggling schools. JROTC takes $50,000 away from each school budget each year and may substitute JROTC courses for other important academic courses (Brown 2011, 138). Additionally, the cost of renovations and upkeep for the schools may place additional fiscal stress on poor schools, and JROTC instruction also costs more than academic, nonmilitary instruction on a per-pupil basis (Clark 2004).

If this were not bad enough, the JROTC program disproportionately enrolls students of color. By enrolling in the JROTC program, students focus on classes such as drill and military science instead of college prep classes or other academic pursuits (Berlowitz and Long 2003, 167). Thus, JROTC actually limits the opportunities for the youth enrolled to prepare for higher education. Additionally, even though the JROTC is said to help "at-risk" students, it rarely accepts students with behavioral problems or low achievement scores (Lutz and Bartlett 1995). JROTC, then, does little to advance the schools and students it purports to help, diverting scarce resources away from poor schools to pursuits that are nonacademic and militarized. What is more, recruitment is a central tenant of the JROTC since the Revitalization Act of 1964 when Robert McNamara (Secretary of Defense) and military officials "redirected JROTC to be more aligned with recruitment and enlistment processes" (Coumbe and Harford 1996, 261). Adding to the allure of JROTC is the fact that the JROTC program is less expensive as a recruitment strategy than other advertising and marketing recruitment programs (Lutz and Bartlett 1995, 10). Thus, not only does JROTC harm schools by gobbling up scarce funding and hiring unqualified instructors, it also channels some of the students most likely to succeed in college into the armed forces, serving as a successful recruitment tool for the US military.

Montgomery GI Bill

Another well-known connection between the military and US education system is the Servicemen's Readjustment Act of 1944, better known as the GI Bill. The GI Bill has been modified several times with each major US conflict (Korean Conflict, Vietnam Era, etc.), but the main purpose behind the Bill has remained essentially the same:

- to help members of the Armed Forces adjust to civilian life after separation from military service
- to give those who cannot afford a higher education the chance to get one
- to restore lost educational opportunities and vocational readjustment to military service members who lost these opportunities as the result of their active military duty
- to promote and assist the All-Volunteer Force program of the Armed Forces
- to aid in the retention of personnel in the Armed Forces
- to enhance nation's competitiveness through the development of a more highly educated and productive workforce (Department of Veteran's Affairs 2005).

Initially, the GI Bill, through the Veteran's Administration, paid a maximum of $500 a year for tuition, books, fees, and other training costs to the educational institution and a stipend of up to $50 a month to participants. In the height of its success in 1947, veterans accounted for an astounding 49 percent of college enrollment (Department of Veteran's Affairs 2005). In 1976, Congress terminated the GI Bill because it was believed to be inefficient and started the Veteran's Educational Assistance Program (VEAP) (Tannen 1987, 50). Unlike the GI Bill, VEAP was a voluntary contribution program that matched individual contributions made by servicemen and women on a two-for-one basis by the DoD with a maximum contribution of $2,700 (Department of Veteran's Affairs). However, with the rising cost of education and failing military recruitment rates, a program of "kickers" was implemented in 1979 under the Multiple Option Recruiting Experiment. These "kickers" were added amounts above the basic VEAP allotments with a maximum benefit of $6,000 (Tannen 1987, 51). The kickers were a success in both cost and recruitment, and in 1982 these added benefits were cemented and added to the basic VEAP benefits under the Army College Fund. The maximum amount payable through these added benefits or kickers rose each consecutive year to keep pace with rising education costs and peaked in 1984 at $18,300 (Tannen 1987, 51). In 1985, VEAP and the Army College Fund were reformed once again and cleverly called the "New GI Bill" and the "New Army College Fund," respectively. Under the New GI Bill, DoD contributions rose to a maximum of $9,600, while contributions by soldiers were capped at $1,200.

One of the biggest changes between 1982 and 1985, however, was that the contributions made by soldiers were no longer voluntary. Any enrollee would automatically have $100 deducted from his or her paycheck for the first year of service. These funds are nonrefundable, and the maximum amount of benefits available under the New College Fund decreased to $14,400 (Department of Veteran's Affairs; Tannen 1987, 52). Although these programs have gone through many modifications, the primary purpose of offering educational benefits to soldiers has remained a central link between education and the military. One of the latent functions of the GI Bill has been to link military service as a pathway to higher education for poor and working-class individuals.

The GI Bill (both new and old) and the Army College Fund are touted as the perfect solution for those who want a postsecondary education but could not afford it. However, the solution fails to address the structural inequalities that have led to inequitable educational outcomes. Notably, those soldiers who benefit from and enroll in these programs are possibly the least in need. Examining enrollment rates prior to the change in 1985 that made contributions mandatory shows the inequity of these programs. From 1981 to 1984, White soldiers were the most likely to enroll in these programs by a wide margin, followed by "Other" racial groups (primarily Asian), then Latino/as, and ending with Blacks as the least likely to enroll in these programs. However, all three racial/ethnic groups have increased their enrollment over the past few years, but the gap between White enrollment and other students is still large (Tannen 1987, 59). For example, even during the highest enrollment rates for Blacks, 20.4 percent in 1983 and 1984, Black enrollment rates were still half that of Whites—40.0 percent. Similarly, the "Other" racial category was 29.5 percent, and the Latino/a rate was 26.3 percent compared to Whites' 40.0 percent (Tannen 1987, 60).

Although this evidence is for the programs prior to 1985 when soldier contributions were mandatory, it does highlight a probable trend among servicemen and women. If less than half of Blacks and Latino/as took advantage of the College Fund when it was on a voluntary basis, how many actually cashed in on the benefits deducted from their monthly paychecks? A study commissioned by the Department of Veteran's Affairs in 2000 found that of those who enlisted in 1994 only 15 percent had utilized Montgomery GI Bill (MGIB) benefits as of 1999 (Klemm Analysis Group 2000, 23). The report additionally found that nonveterans had a higher college degree completion rate compared to veterans, 29 percent versus 21 percent respectively (Klemm Analysis Group 2000, 19). Thus, the GI Bill and the Army College Fund are inequitable across racial and class lines and help those who possibly need it the least. Additionally, as the study commissioned by the Department of Veteran's Affairs illustrated, military service may actually impede the completion of a college degree, and with more and more people of color filling the ranks of the US military, this may serve as another obstacle to obtain higher education for a group already disenfranchised by the US education system.

Troops to Teachers

A more recent collaboration between the military and public education system is the DoD's Troops to Teachers (TTT) program established in 1994 and renewed in 2002 with NCLB. The TTT program places retired military personnel into "high-need" and "low-income" schools as instructors. The primary objective of TTT as posted on their website is to "help recruit quality teachers for schools that serve low-income families throughout America" while at the same time assisting "eligible military personnel to transition to a new career as public school teachers in 'high-need' schools" (Troops to Teachers 2014). State TTT offices offer assistance with certification requirements, employment leads, $5,000 stipends to help pay for certification costs, and bonuses of $10,000 to teach in schools with a high percentage of low-income students (US Department of Education 2008). The schools themselves also receive financial grants to both help with funding and encourage the school to hire military veterans (Anderson 1998, 342).

The TTT program has a negative impact on public education. First, many of the participants in the TTT program have no prior teaching experience and are pushed through the credentialing process (Saltman 2000, 91). Almost 40 percent of TTT teachers have 5 or less years of teaching experience compared to only 12 percent of teachers nationwide. Additionally, over 50 percent of regular teachers have 15 or more years of education compared to less than 5 percent for TTT teachers (Feistritzer 2005, 17). Second, the money funneled into the TTT program to push retired military personnel through the credentialing process and into "high-need" schools could arguably be used in more effective ways, for example, raising the salaries of current teachers to keep experienced teachers working in inner city schools and better funding for educational programs and school infrastructure. Third, the TTT program focuses on "at-risk" students and "high-need" schools, essentially urban public schools with large populations of poor and working-class students of color. Thus, poor kids get discipline-heavy retired military personnel, once again reinforcing the idea that failing schools and students who attend them only suffer from a lack of discipline. However, TTT increases much needed racial diversity for elementary and secondary teachers as 37 percent of TTT instructors are of color compared to 15 percent of K-12 public teaching force. Additionally, 23 percent of TTT teachers are Black and 9 percent Latino compared to 6 and 4 percent in the public teaching force, respectively (Feistritzer 2005, 7). Although TTT increases racial diversity among teachers, it is not evenly distributed between the sexes as 82 percent of TTT teachers are male (Feistritzer 2005, 6). TTT is also part of the deprofessionalization of public school teachers as it follows scripted and narrow curriculum, takes attention away from teacher preparation and pedagogical training and places inexperienced staff in difficult and demanding school environments (Milner 2013; Ozga and Busher 1995).

Federal Development, Relief and Education of Alien Minors Act

The Federal Development, Relief and Education for Alien Minors (DREAM) Act is an immigration bill that, if passed, would allow undocumented immigrants a pathway toward permanent residency and ultimately citizenship. Although different versions of this Bill have been introduced to the US Congress (2003, 2005, 2007, 2011), none have been passed. While the Bill is being lauded as a much-needed piece of federal immigration reform that could potentially benefit 65,000 undocumented immigrant students each year (Amrhein 2009), it is not without flaws. First, many states[1] require immigrants to pay expensive out-of-state tuition, and immigrants do not currently qualify for public scholarships or student loans (Amrheim 2009). The 2011 version of the law would repeal the section 505 of the Illegal Immigration Reform and Immigrant Responsibility Act of 1996 that penalizes states from providing in-state tuition regardless of immigration status. Thus, college-bound undocumented students and their families must pay for college tuition out of pocket, a feat that most undocumented immigrant families are unable to achieve due to the available labor market choices for undocumented immigrants and their families (see Hondagneu-Sotelo 2001). Second, this pushes those undocumented immigrants who cannot afford to pay for college toward the second option of at least two years of military service. While two years may seem a short service to the United States in return for permanent residency, it is misleading, as first-time military contracts are no less than eight years, not to mention the obvious risks associated with service in the US military. If the DREAM Act were to ever be passed, it would substantially increase the available pool of possible recruits to the US Armed Forces and would become another state-supported recruitment strategy linked with education, or more specifically the lack of affordable education, that specifically targets low-income people of color as 80 percent of undocumented immigrants in the United States are from Mexico and Latin America (Passell et al. 2004). In fact, the DREAM Act was included in the DoD 2010–2012 Strategic Plan for military recruitment.[2]

DoD STARBASE and STARBASE 2.0

The DoD STARBASE is a program that invites fifth graders to military bases to learn Science, Technology, Engineering, and Math (STEM) curriculum. According to the STARBASE website, students are provided with 20 to 25 hours of STEM instruction and real-world application (www.starbase.org). They interact with military personnel, explore careers, and, according to the mission statement, are exposed to "the technological environments and positive role models found on military bases and installations." The program specifically targets students "who are socio-economically disadvantaged, low in academic performance or have a disability" and is funded through DoD recruitment funds. DoD STARBASE began in 1993 with seven pilot programs, but has grown to over 76 programs in 40 states, Washington, DC, and Puerto Rico, and since its inception, over 750,000 students have participated in the program (2012 STARBASE Annual Report).

In 2010, the DoD piloted STARBASE 2.0, an afterschool STEM and mentoring program for at-risk sixth to eighth graders. The program specifically targets those students that have previously participated in STARBASE (STARBASE, STARBASE 2.0 Program Overview/Fact Sheet). In 2013, there were 37 middle schools that sponsored a STARBASE 2.0 program. The DoD collects statistics on military attitude by administering pre- and post-tests to students who participate in the program with questions such as "The military is a good place to work" and "Military bases are exciting." In the 2012 STARBASE Annual Report, "military attitude" was one of the attitudinal items that improved significantly from pre- to post-test. The goal of both of these programs is to inculcate students with a positive attitude toward the military, military bases, military personnel, and military careers. The second goal is the militarization of STEM fields which is vital to the military as it depends on them for innovations in modern weaponry and surveillance technology as illustrated by the growing number of Pentagaon-funded research and programs at US universities (Giroux 2007; Gusterson 2009).

What is disturbing about these programs is the fact that not only do the programs focus on poor youth and youth of color as well as the underfunded, struggling schools they attend, but the programs target children who are 10 to 11 years old. If a child attends STARBASE, participates in STARBASE 2.0, and then joins a JROTC in high school, they will have participated in militarized education programs for eight years. Both STARBASE and STARBASE 2.0 are nothing more than recruitment tactics.

Zero-Tolerance Policies

Zero-tolerance policies are disciplinary actions that are automatic and often quite severe and carried out through predetermined punishments and procedures. Under zero-tolerance policies, punishments for school violations are not left up to administrators or teachers, nor are mitigating circumstances taken into account. The clear-cut nature of zero-tolerance policies has led to some quite bizarre and severe punishments such as arrests and referral to criminal courts for seemingly normal behavior by students such as rough housing, use of imaginary guns, or kissing (Rimer 2004; Whitehead 2013). Zero-tolerance policies became popular in the 1990s as a response to rising school violence, but were federally mandated in the Gun-Free Schools Act of 1994, which required schools, in exchange for federal funds, to put in place zero-tolerance policies for any student who brought a weapon to school. However, as schools drafted zero-tolerance policies, many schools included acts and behaviors beyond weapons, including alcohol, tobacco, drug use, sexual harassment, and other types of school infractions (Simon 2006).

Regardless of the beginnings of zero-tolerance, it must be understood within the context of discipline, security, surveillance, as well as the criminalization and militarization of youth, particularly poor youth of color. Schools, especially schools that serve poor youth of color, are understood to be inherently unsafe places. Zero-tolerance policies are thought to make and keep schools safe and work in conjunction with racist and classists assumptions that poor students and students of color are inherently violent (Skiba et al. 2006). Zero-tolerance policies link school infractions with arrests and the juvenile criminal justice system. Police officers are now a common sight at public schools as are metal detector and surveillance cameras. DeVoe et al. (2005) found that the presence of metal detectors in schools was positively related to the enrollment of students of color, and Hirschfield (2009) found that while minority schools represent only about 15 percent of middle and high schools, they represent 75 percent of schools with metal detectors. As a result of such punitive policies and racist ideologies, students of color are more likely to be suspended than White students, and this is particularly true for Black students (Losen and Skiba 2010; Raffaele-Mendez and Knoff 2003; Wald and Losen 2003). Reynolds et al. (2008) examined 20 years of data on zero-tolerance policies and found no evidence that zero-tolerance improves the safety or climate of schools. Thus, zero-tolerance polices are an outcome of neoliberalism and militarization and construct a raced and classed understanding of youth in schools.

Zero-tolerance policies also emphasize individualism and competition through exclusionary practices such as arrest and expulsions. Rather than focusing on democratically defined reforms or solutions to education, such policies move the emphasis from structural inequalities within public education to personal and behavioral issues. The construction of poor students of color as violent, dangerous, and in need of state control functions in conjunction with the neoliberal punitive state, the attack on democracy and the increasing militarization of public schools through the disciplinary control of surveillance. Robbins (2008b) has dubbed this the security-prison-military-industrial complex and argues that such a nexus of neoliberalism and militarization prepares youth for either "a prison order or a social order defined by a perpetual state of war" (89). Giroux (2009) has gone so far to argue that there is a war being waged on poor youth of color under neoliberalism and the "punishing state" views such youth as "disposable" populations.

In effect, each of these programs—zero-tolerance policies, DREAM Act, DoD STARBASE, STARBASE 2.0, TTT, GI Bill, and JROTC—defines poor youth of color as dangerous and failing schools as in need of free-market discipline. Although the United States is, since its inception, a military state, it is important to remember that neoliberal logic facilitates the militarization of public education. Militarized education reforms such as these are part of a historical and imperial project that militarizes public education to control citizens who threaten neoliberal freedom. Under the watchful gaze of neoliberal reformers, failing schools and poor and working-class youth of color are disciplined with free-market ideals and the strict cadence of a military march.

The overwhelming reasons for the overrepresentation of people of color (both domestically and globally) within the US military are poverty and joblessness that many minority youth experience, increasing militarization present in public schools (JROTC, TTT, zero-tolerance policies), as well as the promise of money for college, job skills, and training made by recruitment officers.[3] Thus, poor youth of color become easy targets for military enlistment because of social inequalities, structural and institutional racism, and lack of choice and opportunity in their lives, all of which are exacerbated by the dismantling of the welfare state and the rise of neoliberalism as a political, economic and cultural project.

However, does the military's promise of educational opportunity and employment hold true for enlistees? Barley (1998) compared and contrasted 14 different studies on the effects of military service on civilian earnings from 1976 to 1990. He found that prior to the Vietnam War, veterans of any race or ethnicity benefitted economically from their service in the military. The Vietnam War, however, was a turning point as military service negatively affected the average earnings of Whites, while results for enlistees of color were a mix of positive, negative, as well as neutral affects. However, even with these mixed results, Barley argues that the military is good for people of color because they will receive more training and education then their nonenlisted peers (1998, 153). Barley's argument is weak, however, as he does not specify the types of training or education enlistees will receive, the overall low level of education and training people of color receive in the military compared to Whites, the transferability of these skills to the civilian workplace, or a critical examination of the reasons why so many people of color lack education and job opportunities. Barley suggests that "veterans trained in technical specialties related to computers, electronics, and the repair of electrical and mechanical equipment constitute a...group that benefits from military service because their skills transfer readily to the civilian economy" (1998, 154). What he forgot to state is who is in fact receiving these skills and the risks associated with military service.

While Barley's study is dated the crtiques drawn from the study have not significantly changed. People of color are more likely to fill the bottom-most ranks and less likely to be officers (DoD 2012b). A little over 12 percent of adult homeless populations are veterans (Henry et al. 2013), and veterans in general are more likely to be homeless (Perl 2013). The percentage of all suicides attributed to veterans hovered between 20 and 25 percent between 1999 and 2010 (Kemp and Bossarte 2012), and 22 veterans commit suicide every day (Basu 2013). In 2011, 26,000 active duty members were sexually assaulted (Steinhauer 2013). The Bureau of Justice Statistics estimates that veterans that served during wartime account for 54 percent of state prisoners and 64 percent of federal prisoners (Noonan and Mumola 2007). Additionally, military experience does not translate directly to the civilian labor market (Angrist 1990; Bryant et al. 1993; Goldberg and Warner 1987; Hirsch and Mehay 2003), and veterans are more likely to be unemployed (Humensky et al. 2013; White House 2012). Additionally, veterans are less likely to complete a bachelorette degree than nonveterans (Barry 2013; Department of Veteran's Affairs 2011). Thus, those most vulnerable to the free-market and the destruction of the welfare sate are the least likely to benefit from military service. Military service is a shell game where the promise of jobs, training, and educational achievement is whisked away with a slight of hand.

Thus, because of the global neoliberal policies of the United States, which ensure the predominance of US corporate interests through increasing militarization in the form of invasions and occupations (Cabezas et al. 2007), the US government and military are constantly in search of young bodies and minds with which to carry out its policies and commitments. For example, a study released by the RAND Corporation outlines strategies to increase recruitment of youth who plan on attending college immediately following high school- a group the report referrs to in neoliberal terms as the "youth in the college market" (Kilburn and Asch 2003). Addtionally, the 2004-2005 *School Recruiting Program Handbook* put out by the US Army Recruiting Command illustrates the intersection of education, militarization, and free-market rhetoric. Phrases such as "school ownership," "maximum number of quality enlistments," "sales process," and "total market penetration" jump from the pamphlet that instructs recruiters how to trick kids out of college and into the military through the magic rings of neoliberalism. Neoliberalism, however, does not exist solely outside the borders of the United States because the same logic of competition and accountability that has opened countries' markets has now opened the US public education system. The increasing militarization of the US public education system needs to be understood in relation to the growing influence of neoliberalism and "the enforcement of global corporate imperatives as they expand markets through the material and symbolic violence of war and education" (Saltman 2003, 1). Similarly, Giroux (2004) argues that "neoliberalism has become complicit with this transformation of the democratic state into a national security state that repeatedly uses its military and political power to develop a daunting police state and military-prison-education-industrial complex" (xvii). As mentioned earlier, the turn in public education toward neoliberalism began in the 1950s with Milton Friedman but exploded with the

1983 report, *A Nation at Risk*, that resulted in the ensuing crisis of education, a crisis that opened the doors of public schools to neoliberal reforms and privatization framed in militarized terms. The Pentagon specifically targets schools most likely to yield recruits through militarized programs and policies (Houppert 2005; Medina 2007). That is, the military targets schools located in poor areas with students of color with very little educational or job opportunities (Savage 2004). This is all in contradiction, however, to the Child Soldiers Prevention Act passed by Congress in 2008. The aim of the law was to prevent arms trade with countries that recruit child soldiers. Unfortunately, the United States does not hold itself to the same standard utilizing public education to recruit the most vulnerable youth.

Conclusion

The US public education system has evolved throughout the years. It was initially a loose collection of rural and locally controlled one-room schools, but with the rise of free-market capitalism, educators and leaders looked to business corporations as models for consolidating and gaining control over the fragmented education system. In the beginning years of the common school, education was seen as a way to mold the diverse immigrant populations into US citizens with common values and beliefs. As capitalism became more influential, the common school mirrored the occupational hierarchy of the market, preparing students for jobs. This shift in educational goals moved education out of the realm of the public good and solidly over to a good to be privately consumed and privately chosen. Throughout all these changes in the US education system, one thing that has not changed is the role education has played in the mythical American dream and social mobility.

US citizens hold dear the belief that equal opportunity is offered in public education, but this dream was shattered in the 1980s as reports out of Washington touted the failures of the "one best system." A flurry of academic and policy-related work suggested ways to fix our failing schools, and with Thatcherism and Reaganomics at the helm, educational reform made a right turn toward neoliberalism. Neoliberalism affects how we trade, work, live, shop, and now learn. The magic hand of the market is held up as the holy grail of educational reform from Friedman, Chubb, and Moe to charter schools and NCLB. However, neoliberal policies that invoke "market solutions" such as competition, choice, accountability, and efficiency are not improving schools. Neoliberal policies actually reproduce the social inequalities and hierarchies that are prevalent in the US education system by pushing more and more poor students of color away from higher education and toward the military. This makes public schools more vulnerable to militarization as educational programs and funding are held as hostages in exchange for access to US youth. This results in the most vulnerable youth being under attack for military recruitment and pushed by neoliberal practices to barter their educational opportunities with military service. Since, the 1970s, the US military relies on recruitment to fill the ranks of the all-volunteer force and utilizes legal means, advertising, and public schools to garner access to US youth. Militarized educational programs and policies such as the JROTC, the GI Bill(s), TTT, the DREAM Act, DoD STARBASE, and zero-tolerance policies work to recruit poor and working-class youth of color who are marginalized by the capitalist, neoliberal market. Neoliberal policies and militarization are forces that work together and implemented within public schools located in poor and/or working-class neighborhoods and communities of color. The spread of neoliberalism and militarization, the twin forces of US imperialism, has made equal opportunity in US public education a distant American dream for these youth.

3

Sending Good Kids to Military School: Why Parents Choose the MEI?

> First impressions are huge, okay, and if you have parents that are together, okay, and one parent has a kid at the MEI walking toward that parent group; of course they are going to smile and be proud because it is a status issue. It is almost like private school…almost. So, there is a huge status thing and the parents are proud that their kids are in this school and yeah, there is a huge status thing.
>
> —Mr Thomas, an MEI teacher

As outlined in the previous chapters, charter schools and neoliberal restructuring of public education are profoundly racialized. Charter schools are presented as the neoliberal solution to "failing" public schools and the educational gap, which are both overwhelmingly concentrated in poor communities of color. Additionally, neoliberal educational reforms intersect with militarized educational reforms and programs (e.g., JROTC, on-campus police, security and surveillance technologies, TTT program, etc.) to construct a web of Foucauldian (1975) disciplinary measures and control utilizing the assumption that all struggling schools require militarized discipline in order to succeed. However, this approach ignores that struggling schools and the children they enroll are "failing" due to historical economic and educational inequality, not due to a lack of market or militarized discipline.

Why are charter schools widely accepted as a solution to US public education if they do not actually perform at a higher level than noncharter public schools and are deeply racialized and inequitable? Why do parents send their children to schools that may be no better than regular public schools, and in some cases worse? Studies suggest that parents enroll their children in schools for a variety of reasons, not based solely on school performance, including the race and class composition of the school (Schneider and Buckley 2002; Weiher and Tedin 2002), transportation costs and proximity to home (Andre-Bechely 2007; Goldhaber et al. 2005). While this may hold true for parents with a variety of educational options or resources with which to choose (information, time, income), what happens when parents are marginalized by racist or classist public education systems and excluded from the democratic participation processes? What happens to parents who have limited resources and hence choices? This chapter will examine these questions in regard to parents and students in the community of Eastmoore, specifically examining the motives of parents to enroll their children in a militarized charter school. The chapter begins with a discussion of neoliberal discourse of freedom that, through the Gramscian notion of consent, leads marginalized and oppressed parents and students to participate willingly in neoliberal educational reforms and, in this case, militarized educational reform. This general discussion is followed by the examination of the specific community and educational contexts that allowed for the founding of a militarized charter school in Eastmoore. Ultimately, I argue that the lack of alternative equitable education in an overcrowded, underfunded, and academically struggling school district, combined with racial tension, above-average test scores, attraction of a school uniform, and perceived need for disciplinary education, propels parents to enroll their children at the MEI.

Consent, Choice, and Illusions of Empowerment

Freedom as an ideal is held dear in the United States and taken as an unquestioned fundamental ethos of the nation and the foundational discourse of neoliberalism, which argues that individual freedoms are upheld through the freedom of the market. This is a powerful ideology that shapes much of how the United States interacts not only with the world but also how the United States and its citizens understand domestic policies and reforms. Freedom as a fundamental value is difficult to argue against because it is deeply embedded in the cultural and political underpinnings of the United States. Who would be against freedom whether it is individual freedom or market freedom, or the conflation of both? Thus, freedom is so engraved in our cultural and political traditions that it has become what Gramsci (1971) defines as "common sense"—that is, a commonly held belief by a group of people. However, "common sense," according to Gramsci, forms the basis for "consented coercion" and is produced and reproduced through the hegemony of the dominant class. That is, the interests of the ruling class are accepted in society as being interests of everyone through a network of ideas, institutions, and social relations. This is easily illustrated with the neoliberal construction of free markets and choice as ensuring individual freedoms, (which in reality are beneficial only to the ruling class and the business and political leaders of the society) while usurping real democratic power and participation in society.

Neoliberalism is a powerful pedagogy spreading the logic of the market. Its implementation may have begun from the top down, but it becomes a part of everyday practices and understandings of marginalized and oppressed individuals. This is particularly helpful in understanding why parents and communities support neoliberal or militarized educational reforms and policies. It is important to understand that such Gramscian consent occurs within highly racist and classed educational contexts. For example, Pedroni (2007) found that the voucher program resonated with African-American parents in Milwaukee. as it offers these parents dignity and agency as they identify themselves as educational consumers rather than dependent citizens reliant on a racist and classist public education system. The emancipatory potential for self-determination and the meaningful education of marginalized groups is demonstrated in other neoliberal reforms such as magnet and charter schools (Rofes and Stulberg 2007; Stulberg 2008; Wells et al. 2002). Thus, neoliberal education reforms can offer marginalized individuals and communities the promise of hope, self-determination, and agency. They offer an alternative to dominant school practices and structures that are historically and economically inequitable. However, the identification of marginalized groups with neoliberal educational reforms points to a lack of social and political power rather than agency. It highlights the "sociopolitical debt" (Ladson-Billings 2006) and marginalization of these communities from the civic, democratic, and decision-making processes that ensure equitable and quality education. Parents at the MEI, as I argue in this chapter, align themselves with neoliberal reforms and consent to the "common sense" of militarized choice due to the educational, historical, and social inequalities of their lives and the lives of their children, specifically the underfunded, overcrowded, and violent Eastmoore School District.

The Making of a Militarized Charter School in California

California was the second state to pass charter school legislation (1992) and houses the largest raw number of charter schools in the country (US

Department of Education 2013c). Charter schools are public schools paid for with tax dollars, but are exempt from most of the state laws governing school districts. California charter schools cannot charge tuition and are open to all students within a school's county or adjacent county. The process of founding a charter school involves petitioning the governing board of the school district and drawing up a "charter" or school constitution that addresses required elements. The required elements of a petition include, but are not limited to, a description of the educational program of the school, governance structure of the school, measurable student outcomes, targeted student population, and admission requirements if applicable (California Education Code: 47605(5A-P)). In addition, financial statements must also be submitted that include a proposed first-year operational budget, start-up costs, and cash flow and financial projections for the first three years of operation (California Education Code: 47605(G)). Parents or guardians must sign the petition, and the number of signatures must represent at least half the number of students estimated to enroll in the charter school in the first year. Teachers must also sign the petitions, and the number of required signatures is at least half of the number of estimated employed teachers in the first year. The governing school board then has 30 days to hold a public hearing on the charter after which another 60 days are allowed to either grant or deny the petition (California Education Code 47605(B)).

As outlined in chapter 1, charter schools, like the MEI, are predominantly located in poor or working-class communities of color and primarily enroll students from the surrounding communities (Manno et al. 1999; Nelson et al. 2000). More than 83 percent of the community of Eastmoore identifies as belonging to a racial or ethnic category other than White. Eastmoore is also overwhelmingly working-class; the median household income is $36,000 compared to the state median of $57,000. Of the seven schools in the Eastmoore School District, three do not qualify for Title I funds[1]: the MEI, an online charter school, and a high school located 10 miles southeast of Eastmoore. For students looking to attend a school that is not at risk for academic failure, or for those students who lack self-motivation for online learning and transportation to a school outside of Eastmoore, this leaves the MEI as the only available public school option. In addition, until the opening of the MEI in the 2002–2003 school year, the school district had only one middle school. Eastmoore Middle School enrolls over 1,300 students and operates on a year-round schedule with four different calendar tracks due to overcrowding. Overcrowding and a continuing surging enrollment is a problem at all of the Eastmoore schools and served as a catalyst for a $46 million bond approved by district voters in November 2004, designed to ease the bulging of schools and classrooms. The rising enrollment numbers at Eastmoore Middle School, as well as the rising enrollment numbers of other schools in the district, is one of the justifications behind the founding of the MEI. The MEI has expanded one grade each year since it opened in 2003, fully incorporating seventh through twelfth grades, alleviating some of the overcrowding in the Eastmoore School District, particularly in middle school grades.

Charter schools place the responsibility for developing a curriculum, hiring teachers and staff, as well as other administrative duties on the shoulders of the oversight committee and the organization managing the charter school. As a result, the MEI is a quick and easy fix to the Eastmoore School District's problem of overcrowded schools. It is a solution that requires minimal organizational support from the Eastmoore School District. It is also a solution couched in the neoliberal rhetoric of choice. Prior to the opening of the MEI, poor and working-class parents had only one option for a middle school, one that was literally bulging at the seams. Mr Jones, a teacher who has been with the MEI since its opening, related in neoliberal terms how the Eastmoore School District had an interest in the opening of the MEI due to the overcrowded conditions of the local middle school and the lack of choice for parents and students in the district. "The middle school is overcrowded, at capacity. Eastmoore School District had a vested interest in MEI... [the] need to give parents a choice."

Although the Eastmoore School District was in desperate need of another middle school due to overcrowding, why did the school district support the opening of a militarized charter school? The school district had the option to open an additional public middle school or even another type of charter school, why then was MEI the solution? An integral part of the answer to these questions lies in neoliberal discourse and the social context of the Eastmoore School District. The Eastmoore School District is plagued with violence and racial tension. Around the time of the opening of the MEI, a high school in the area was locked down after an argument between two students incited a race-based fight involving 500 students that resulted in three arrests and the detention of over 20 students. Additionally, Eastmoore High School was locked down after a similar racially motivated incident. During my first year at the MEI, there were six racial incidents in Eastmoore schools.

Žižek (2008) argues that illustrations of subjective violence are the result of systemic violence such as class, race, gender, and sexual oppression and is used to maintain domination and exploitation within society. Thus, individual acts of violence, such as the case in Eastmoore schools, can be attributed to the structural violence of inequality in the local community in terms of racial oppression and class injustice. However, violence is not viewed as an indicator of social inequality, but rather of dangerous and uncontrollable youth, especially young men of color. As Robbins (2008a) has pointed out, the solution to such subjective, individual violence is the militarization of schools and school environments in order to support a "system of social control used to respond to the 'emergency' of school violence" (342). Thus, the militarized format of the MEI took precedence over other forms as a solution to violence in the district schools and to build good kids through militarized discipline.

Academics at the MEI

The MEI is marketed to prospective parents and students as a school that focuses on academics and prepares students for success in secondary education. The school sells itself as "college prep," confidently advertising this on recruitment fliers and informational packets. Ironically, the MEI downplays that fact that the school is fully militarized. The recruitment flyer for the school states that while the school has "a military format and class structure" it does not have a "'boot camp' type environment." The bold claims of academic excellence are peculiar, considering this information was printed on a recruitment flier and distributed to the community prior to the school opening and evidence of its academic outcomes. Are the claims made by the MEI administration and Eastmoore School District valid?

First, although the MEI relieved some of the overcrowding of the local middle school, there is an enrollment capacity due to the physical limitations of space and limited funding for teachers and support staff. This allows the MEI to control which students are admitted and to allow them to screen out low-performing and/or high-needs students. The MEI has a list of eight enrollment requirements, only two of which (numbers 1 and 2) specifically focus on academics:

1. Complete last grade level with a "C" average (2.0).
2. No failing marks in any core subject area earned during the most recent recorded academic grading period. STAR[2] testing results for the previous year must show proficiency at the "Basic" level or higher.
3. Applicants must not have ever been expelled from any school or school district for any reason.
4. Applicants must provide documentation of a discipline record free of any suspensions within the previous school year and without any pattern of disruptive or defiant behavior.

5. Have no more than 10 absences in past school year.
6. Agree to abide by Cadet Code of Conduct.
7. Compete in Presidential Physical Fitness competition.
8. Be willing to:
9. be assessed for grade-level skills
10. be interviewed
11. write application letter
12. wear a complete MEI uniform daily.

MEI boasts that the school offers a "college prep" curriculum, but the evidence suggests otherwise. For example, a 2.0 grade average with no failing grades in the previous year is hardly a rigorous academic requirement or a standard for college admission or preparation. The MEI also fell short in other ways, especially computer access, limited curriculum, and inexperienced and emergency credentialed teachers. During the first two years of operation, the school had no computer access for the students and still lacks access to a library. In fact, the school did not have computer access until the fifth year of operation, and access was limited and up to the discretion of the teacher. At that time, there was no computer lab or physical space in which students could access the laptops. The laptops were stored on a large cart in the broom closet and wheeled into classrooms when needed, usually so that cadets could complete online research or as a reward for meeting learning and/or behavioral goals. A school with limited computer access is more consistent with public schools in the inner city than it is of a college preparatory school, particularly considering the centrality of computers in our daily and work lives and the importance of computer and internet use pedagogically (Valadez and Durán 2007).

Second, the MEI has a very limited curriculum. The MEI had a "no homework" rule in the first two years because there were not enough textbooks for every student to take home. Until the sixth year of operation, it offered only the basic classes in math, science, and reading. There were no advanced placement classes offered at the MEI, no honors courses, and during my time of observation, no electives. The MEI also lacked accreditation by the Western Association of Schools and Colleges for the first five years of operation.

During my time at the MEI, only 10 (66.7 percent) of the 15 teachers were fully credentialed. The remaining teachers were either interns or possessed emergency teaching permits.[3] As shown in Table 3.1, the teachers at the MEI have a lower credential rate and years of experience compared to the teachers in the local district and the state. The percentage of fully credentialed teachers in the school district is 89.5 percent, and in the

Table 3.1 Teacher Credential Rates and Experience

	Full Credential	Emergency Permit	Average Years Teaching
MEI	10 (66.7%)	3 (20.0%)	3.8
District	342 (89.5%)	13 (3.4%)	11.5
State	294,898 (95.0%)	1157 (.4%)	12.8

state 95 percent. MEI also has a higher percentage of emergency permit holders than in the district (3.4 percent) and state (0.4 percent). In addition, according to the California Department of Education website, the average years of teaching experience at the MEI is 3.8, noticeably lower than in the Eastmoore School District (11.5 years) and state (12.8 years).

Multiple studies have illustrated the importance of teacher experience in regard to student performance and outcomes. Darling-Hammond and Youngs (2002) found that student-teacher experience is an important factor in teaching effectiveness, while Rice (2003) and Goe (2007) found teacher experience to be particularly important in the first years of teaching. Thus, hiring inexperienced teachers is a detriment to the academic climate of the academic success of the MEI and its students.

Although the MEI has branded itself as an academically focused charter school, the school's lack of experienced and fully credentialed teaching staff, limited curriculum, and lack of or limited access to computers stand as glaring contradictions to the claims of academic excellence. With that in mind, how does the academic performance of MEI cadets actually compare to students in the Eastmoore School District and across the state? One of the most commonly used measures for identifying academically successful schools is standardized test scores. Although standardized tests have been highly criticized (Crouse and Trusheim 1988; Fleming and Garcia 1998; Goslin 1967; Jencks and Philips 1998; Neill et al. 2004; Taylor 2002), parents and administrators still view these as credible indicators of quality education and good schools.

In terms of standardized test scores, MEI students perform better than students at the district and state levels. For example, the California High School Exit Exam (CAHSEE) is a test all students in California must pass in order to graduate from high school. During my time at the MEI, students had a 90 percent passing rate for the English language arts section of the test and a 95 percent passing rate for the math portion (see Table 3.2). In contrast, scores at the district level were 59 percent for English language arts and 56 percent for math, and 79 and 78 percent, respectively, at the state level (California Department of Education).

Table 3.2 CAHSEE Passing Rates for MEI, District, and State

	English Language Arts	Math
MEI	90%	95%
District	59%	56%
State	79%	78%

Table 3.3 CST Scores for English Language Arts and Math*

	MEI (Language Arts/Math)	District (Language Arts/Math)	State (Language Arts/Math)
Grade 7	49% / 35%	27% / 27%	49% / 41%
Grade 8	48% / 27%	30% / 14%	45% / 18%
Grade 9	44% / –	41% / –	49% / –
Grade 10	53% / –	34% / –	41% / –
Grade 11	45% / –	31% / –	37% / –

*Percentage of students "at or above proficient."

The California Standards Test[4] (CST) measures how well students master specific skills by grade level. The CST grades students as "proficient or above." Table 3.3 shows that MEI students scored above the school district average for language arts in grades 7 to 11. With the exception of the ninth grade, MEI students scored at or above the state average in language arts. For math, seventh- and eighth-grade MEI students scored higher than the district average but had mixed result in comparison to state test scores (California Department of Education).

In general, while MEI test scores may not be exceptional, they do surpass other district schools. MEI students scored above the other middle school and high schools in the Eastmoore School District. This allows the MEI to be seen by the Eastmoore community, despite its multiple inadequacies (e.g., varied test scores, limited curriculum, inexperienced and noncredentialed teachers, lack of books, library, and computer access) as an academically successful school. However, the test scores of the MEI need to be closely examined. First, the MEI does not accept students with lower than a 2.0 grade point average or students who do not pass at a "basic" level on statewide standardized tests. If cadets fail to meet any of these requirements during the course of the school year, that cadet may be removed. Second, the MEI does not serve any special needs or ESL students whose relative educational challenges may decrease the MEI's results on standardized tests. Thus, the MEI has a "creaming effect" upon the local school district, leaving the middle school and other high schools in the school district to enroll the neediest students. The MEI is able to maintain its higher test scores then by not accepting underperforming or at-risk students, only accepting the top applicants for enrollment, and dismissing students who do not adhere to the behavioral or academic standards of the school. These practices bolster the MEI's test scores and enhance the school's reputation in the Eastmoore community. Claudia, an outgoing Latina who has the highest grade point average in the eleventh grade, explained to me: "All the schools [in the district] here are not very advanced. So, MEI sticks out because of that. Because our test scores, like state testing, our CAHSEE." Thus, due to the dismal performance of other schools in the district and the ability to choose which students attend and remove those who do not meet testing standards, the MEI is able to outperform other schools in the local school district. Thus, the MEI is the best possible option within a limited and struggling school district. However, this only takes into account the academic record, or more precisely the testing outcomes of the MEI. What impact does the militarized structure of the school have on the decisions of parents to enroll their children at the MEI?

Sending Good Kids to Military School

Many military schools and programs are marketed as a last resort for troubled teenagers and use strict discipline to set "bad" kids straight. A quick Google search for military schools that serve troubled teens brings up over 100,000 results and thousands of military schools and boot camp programs. The majority of the schools are private and expensive ranging from $9,000 to over $30,000 a year for tuition and related costs- out of reach of most parents and families in Eastmoore. The MEI is different from these schools and programs because it is a public charter school with the only additional cost to attend the purchase of the required MEI uniform.

However, the MEI is not specifically designed for students with discipline problems, and downplays the fact that it is a militarized school. The MEI is marketed as an academic school that preps students for college, something many of the working-class parents of Eastmoore were not able to attain themselves. Major West, the initial commandant of the MEI, repeatedly stressed the centrality of academics and made major efforts to distance the MEI from stereotypical military schools that cater to troubled teens. In my initial meeting with Major West, he stated repeatedly that the MEI is just like the military; it is high-achieving and accepts "only the very best." The recruitment flyer stated, "the [MEI] has a military format and class structure, but does not have a 'boot camp' type environment." In short, the MEI has, in free market terms branded itself as an academic achieving charter school that just happens to be militarized.

Since its inception then, the MEI has been promoted as an academically focused school to the parents and students of Eastmoore. However, as discussed above, the school's academic record does not necessarily match its reputation. If students are not being sent to the MEI because of disciplinary problems, and the academic record of the school is contradictory, what reasons compel parents and students to choose the MEI? There are three primary reasons why parents and students choose the MEI: lack of alternative educational choices, school discipline, and the school uniform.

There Is No Alternative

Parents and students choose to attend the MEI because of a lack of alternative educational choices. This is particularly true for middle school students. The majority of the MEI's enrollment is in grades seven and eight. School enrollment decreases substantially for high school grades nine to twelve. The first graduating class of 2009, for example, consisted of 30 students. Many students who choose to attend the MEI for middle school move on to one of the two high schools in the district that offer a more traditional high school experience in terms of sports and extracurricular activities as well as a variety of elective classes, music, and arts programs. Like most charter schools, the MEI has a very limited curriculum (discussed earlier) due to its small size and limited budget.

There is only one middle school in the Eastmoore School District other than the MEI. There are three main reasons parents and students provide for choosing to attend the MEI over Eastmoore Middle School. First, Eastmoore Middle School has a dismal academic record. During my time at the MEI, the Eastmoore Middle School scored below state averages on all parameters of the CST and had an Academic Performance Index (API) of 1. A school's API is a number that reflects performance level based on statewide tests and ranges from 200 to 1000. A statewide API ranges from 1 to 10. A score of 1 means that Eastmoore Middle School fell into the bottom 10 percent of schools at the same grade levels in the state. The MEI in comparison has a statewide API of 6, meaning that it falls into the 60th percentile for comparable schools in the state of California. In the eyes of many parents and students, Eastmoore Middle School provides poor quality education. As the seventh-grade Social Science and English teacher, Mr Park, told me, "parents talk in the community and I have heard parents talk about [Eastmoore Middle School] they say it is worse." Thus, parents and students consider the MEI to be a better educational opportunity. Two students related how their parents saw the MEI as a "better chance" for them. Zack, a White seventh grader, said: "my parents wanted me to come here so I can get a better chance at going to college." Similarly, Miles, a soft-spoken, Black seventh grader, stated: "my parents wanted me to get an education [at the MEI] because it is a better chance for me. And at the public school, the teachers are in a hurry and they try to hurry up and get us out of school so they can get on with their life and stuff." While I cannot speak to the behavior of teacher across the Eastmoore School District, Miles believes that the teachers at the MEI are more responsive to his and other students' needs and will result in better educational outcomes.

A second reason for choosing the MEI is the overcrowding at Eastmoore Middle School. Eastmoore Middle School enrolls over 1,300 students and runs four different calendar tracks on a year-round schedule. Parents and students perceive that overcrowding leads to disorganization, bureaucracy, and chaos. This discourse is interestingly similar to the neoliberal arguments made against public schools (inefficient, undisciplined). As Steven, a Latino student with the highest military rank at the MEI, stated to me: "Eastmoore is a disaster. I went there four times to sign up and still never did…thought the MEI was better organized and something new to try." Thus, overcrowding and a perceived lack of organization at Eastmoore Middle School prompted parents and students to look for another option.

A third reason parents and students choose to attend the MEI is because of violence and racial tensions in Eastmoore schools. As discussed earlier, Eastmoore School District and the Eastmoore Middle School, in particular, suffer from severe racial tension that has resulted in race-based fights and tensions. The community of Eastmoore and the Eastmoore School District is racially and ethnically diverse. The racial and ethnic distribution of students in Eastmoore schools is as follows: 0.5 percent American Indian or Alaska Native, 1.5 percent Asian, 0.7 percent Pacific Islander, 2.1 percent Filipino, 63.2 percent Latino, 9.6 percent African American, and 22.1 percent White (California Department of Education). The Life Science and World History teacher, Mr Thomas, who had spent ten years in the US Air Force as an operating room medical technician, believes that racial tension is the primary motivation for parents to send their kids to the MEI:

> You hear what is going on in [the area] with the race riots and stuff like that. Parents are very concerned, obviously, because they don't want their kids hurt. So, there is a strong influence of gangs, you know, racial tension, that type of thing. You don't see that here and if you did, then it gets squashed immediately. You know, because we have a smaller student body so we can deal with issues a lot quicker and nip them in the butt [sic]. Parents know that and this is pretty much, I think in my opinion, why they send their kids here.

Susan Uppal, a White parent who is extremely involved with the MEI as a member of its advisory board and volunteer for school functions, field-trips, and parades, expressed a similar concern for her daughter, Cathy, who attended the MEI for seventh and eighth grades:

> Eastmoore has a lot of racial issues. As you know, Eastmoore is 85–90 percent Hispanic. So there are a lot of problems with Eastmoore. There is [sic] a lot of racial problems... going on there. So parents are skeptical to send their kids there. And that was a factor for me because in 6th grade the problems I could see were already starting. I knew that if she went there [Eastmoore Middle School], then, it was probably going to escalate. But her friends say, "Well, we don't have problems here," but they are Black or Hispanic. They are not White. Unfortunately, in this day and age, it makes a difference to these kids.

Susan Uppal's comment is couched in the racist assumption that people of color are dangerous and highlights the complexity of negotiating race for both students and parents in Eastmoore. Although race does not matter to her, it does matter to "these kids" and, she believes, this has consequences for her daughter and her education and safety. Likewise, Estella Garcia, a Latina mother who operates a small hair salon out of her home, said she was worried for her daughter's safety and educational future because of violence and gang presence at Eastmoore Middle School: "I know all my neighbors around here. I said to Claudia, 'you are not going to go there.' I was so scared. I didn't want Claudia to get into something that would affect her education."

Many of the MEI students also voiced concern about Eastmoore Middle School as a violent school. For example, Erika, a Latina eighth grader, told me that the MEI "is a good school compared to Eastmoore, lots of fights at Eastmoore." Similarly, Michael, a Black seventh grader, explained to me: "Eastmoore has gangs and get into fights so I didn't want to go there." Finally, Clarissa and Marilis, both seventh graders and Latina, made similar statements: "nobody's going to beat you up [at the MEI]" and "everyone is nice here, no fights." Students, parents, and teachers agree racial tension and violence are reasons to avoid Eastmoore Middle School. However, the MEI is not free from conflict and fights (discussed in Chapter 5), but the belief that the structure of the MEI and its small student body results in fewer fights and a safer learning environment motivates parents to choose the MEI.

During my observations at the MEI, I did not witness any account of racial animosity or tension between cadets, although they may have occurred. On the other hand, student cliques at the school were racially diverse, and self-segregation by race was minimal. The most racially isolated group was a small group of four to five eighth-grade Black girls who would watch the boys (mostly Black and Latino) play basketball during lunch break. Most Black girls were integrated into racially mixed cliques. However, there were cliques at the MEI that were predominantly Latino, which is to be expected at a school with almost 70 percent Latino enrollment. The racial diversity at the MEI could be due to the racial diversity in the Eastmoore community. It is also possible that the cliques at the MEI were bound by class similarities, but this was hard to discern due to a lack of markers of class status such as cell phones, portable electronic music devices, street clothing, and jewelry, all of which were banned at the MEI. Additionally, most cadets at the MEI lived in the surrounding Eastmoore community that was overwhelmingly working-class, so class stratification may have been slight.

In the end, part of MEI's success is attributed to the failings of Eastmoore Middle School: violence and racial tension, overcrowding, and low academic performance. As Eastmoore Middle School is the only available public school in the area, and most students and their families cannot afford private school, they embraced any type of alternative, even a militarized one. As Susan Uppal stated, "Some of the other kids' parents I've talked to [said] it is kind of the same thing. They didn't want them to go [Eastmoore Middle School] and [the MEI] was their best alternative." Additionally, Mr Park summed up the reasons behind the growth and success of the MEI in a succinct, single phrase: "anything but Eastmoore Middle school." The fear parents and students have of a bad school populated by bad kids is important to understand, as it is a precursor to the second reason behind the success of the MEI—discipline.

Discipline

Although the MEI is not a disciplinary-based school or a military reform school, discipline is one of the most cited reasons parents send their children to the MEI. Parents believe that the militarized discipline and structure of the MEI will enable their children to achieve academic success and avoid gang and drug involvement. Students often shared stories of Eastmoore Middle School that involved drug sales and drug use on campus, violence, and gangs. Ana Lopez, a mother who had come to the MEI every morning for two years to watch her son, Chris, lead the morning formation, appreciates the discipline of MEI: "I wanted my son to come here because of the discipline and to try something different. The MEI is more organized than other schools and shows students how to be more respectful... discipline helps them get structured." Accordingly, Mr Jones stated that the structure of the school positively impacts cadet behavior:

> Here [at the MEI] there is more order. Most of the time you see the kids being much more polite and much more goal oriented. You know, they are here for a reason and their parents brought them here for a reason and their parents want them here for a reason. They want that type of instruction for their kids.

Claudia also distinguished MEI from other public schools: "in [the] MEI if you don't have good grades, if you don't have good attitude, if you don't attend school, I mean you can't do what you can do in a regular high school... just not go. You have to follow rules." Even substitute teachers comment on the behavior of the cadets at the MEI, often leaving their contact information with the front office in hope of being asked to work at the MEI.

Discipline at the MEI is militarized and based upon the Cadet Code of Conduct and the Six General Orders for Discipline, which the administrators and teachers argue leads to the academic and social successes of MEI cadets. Each cadet is given a handbook at the beginning of the school year that outlines discipline at the MEI and the consequences of violating either the Code of Conduct or the Six General Orders. The Code of Conduct is modeled after West Point's code of conduct and states that "no cadet shall lie, cheat, steal nor tolerate those who do." MEI cadets must memorize the Code of Conduct and be able to perfectly recite the code when asked by teachers, staff or administrators. The Six General Orders for Discipline are part of the militarized disciplinary structure of the MEI:

1. I will comply with any instructions or directions given to me by any adult staff member of the [MEI]. I may disagree with the instructions, but I will always comply with them.

2. I will treat all members of the Corps of Cadets,[5] staff, visitors and parents with absolute courtesy and respect. I am not obliged to socialize with any member of the Corps of Cadets, but I am required to be courteous and respectful at all times.
3. I will neither commit nor condone any acts of violence or threats of violence against any member of the Corps of Cadets, faculty or staff for any reason. It is my duty to assist others in resolving conflicts without violence or to seek assistance before any violence is committed.
4. I will respect the property of others and of the [MEI]. I will take care of all property entrusted to me and will return all such property in a timely manner and in the same condition as when I received it.
5. I will never leave the campus of the [MEI] without permission from staff members.
6. I will conduct myself at all times in such a manner as to reflect honor on myself, the Corps of Cadets and the [MEI].

Consequences for violating the any of the school rules are, but are not limited to, the following:

1. Assignment of a physical training regimen to include no more than 10 regular or modified push-ups or assignment of close-order drill for not more than 20 minutes each period of instruction.
2. Assignment to a supervised work to perform routine cleaning and maintenance projects.
3. Detentions before, during, or after school.
4. Assignment to Saturday School.
5. Assignment of written essays or other written work.
6. Exclusion of a cadet from attendance during an investigation or due to a reasonable concern for the safety of the cadet or other cadets.
7. Suspension from school.
8. Dismissal from the Corps of Cadets.
9. Expulsion as provided for in section 48915 of the California Education Code.

The goal for MEI cadets, according to administrators and teachers, is to develop "self-discipline" through the regulations and militarized structure of the MEI. This is done through the strict adherence to rules of behavior, standards of appearance, and timekeeping. The idea of "self-discipline" is important as it removes the disciplinary measures of the school onto the shoulders of the students. That is, control is exerted through the students as not only disciplined objects (school rules) but also as self-examining and disciplinary subjects (self-imposed discipline) (Foucault 1978), a more powerful and subtle measure of control as students internalize the militarized disciple of the institution.

The rigid, rule-based and militarized structure sets the MEI apart from other schools in the area. Mr Park explained to me one afternoon in his classroom: "there is more structure here at MEI, which schools need. Schools in general need more discipline. It is the small things; the attention to detail that is important, it retains students' attention and trains them in discipline." The adherence to rules is a key aspect that attracted Estella Garcia to the MEI. She felt the rule-based structure of the school would prove to be a valuable lesson for her daughter, Claudia.

> A lot of kids are not disciplined and a lot of times they can't follow rules, they can't follow directions and in the real world...everything is based off of a rule. You have to follow the rules and kids don't learn that and then they go out and they start doing things and it is, they don't care. And the way that they teach the students the rules and the way to follow them is what I like. So that [my daughter] knows and understands that there are things that she has to do, even though she might not like them, she has to do them.

Susan Uppal also agrees that the militarized discipline and adherence to rules is important not only for the success of the MEI but also for the future success of the students:

> I think that the structure and the discipline is important and it helps mold who they are...It is all about being structured and learning right from wrong. This is the age when they are going to go one way or the other and junior high defines who they are. So, the military structure for junior high is a very big factor.

Both parents and teachers stress the adherence to rules and discipline. It is not the rules themselves that are important, such as respect for peers or appearance standards, but the strict adherence to rules that takes precedence. The strict adherence to rules and obedience is what Jackson (1968) refers to as the "hidden curriculum" or the unintended and unofficial knowledge that students learn in educational settings in addition to the official curriculum (math, science, reading). The hidden curriculum can include such expectations as proper school behavior (standing in line, raising hands), or the knowledge of how to effectively interact with bureaucracies (forms, records, hierarchical ordered offices, and positions). What is important about the hidden curriculum is that it often reproduces social inequalities through the subtle messages and skills that students receive as it varies across different levels of social stratification such as race, class, and gender. For example, different social behavior is rewarded according to social class. Middle-class students are rewarded for critical thinking and managerial skills, whereas working-class students are rewarded for strict obedience and rule compliance. At the MEI, the hidden curriculum is particularly apt for race and class as the cadets at the MEI are taught strict adherence to rules and obedience through militarized discipline. Bowles and Gintis (1976) argued that this structure of education reflects the structure of the labor market. While Bowles and Gintis are overly deterministic and downplay agency and resistance, the theory does highlight the expectations for students based on the curriculum of the schools in which they attend. Thus, for working-class kids, like those at the MEI, the strict adherence to militarized rules and uniform standards mimic not only working-class labor conditions but also the disciplinary structure of the military and readies cadets for entrance into working-class jobs or enlistment in the armed forces. As Mr Park so bluntly stated:

> [The uniform] also teaches students responsibility, how to be properly dressed, shirt tucked in, make sure your buttons are done all the way up to the top one, and make sure your shoes are shined, make sure your pants are clean. It basically teaches them, you know, how to prepare for a career, prepare for the workforce.

Although Mr Park was referring to the MEI uniform, it was the strict adherence to uniform guidelines and rules that was stressed, not the uniform itself. Mr Park's comments highlight the importance of self-discipline or "responsibility" as MEI cadets will most likely (he assumes) end up in working-class jobs that utilize rigid disciplinary measures of control.

Not only is class an important factor that influences the structure of the MEI, but so too is race as students of color are more likely to be surveillanced and punished because they are perceived by school personnel as needing strict school discipline (Children's Defense Fund 1975; Ferguson 2000; Foucault 1977; Skiba et al. 1992). Thus, as the MEI overwhelmingly enrolls students of color (86.5 percent), the militarized discipline of the school coincides with the argument within neoliberal ideology that poor youth of color are dangerous, violent, and in need of state and militarized discipline. That is, the militarized discipline at the MEI is viewed by parents, teachers, and administrators as the best way to educate and control Eastmoore youth.

In sum, the MEI attracts parents because of its militarized discipline structure even though it is not specifically designed for troubled teens or students with disciplinary problems. The militarized structure of the MEI and the strict adherence to rules and regulations is part of the hidden curriculum preparing working-class kids for working-class and enlisted lives controlled through the disciplinary measures of the surveilling neoliberal state. Moreover, as the Eastmoore School District is violent and academically struggling, the militarized discipline offered by the MEI is viewed as a way to set troubled kids straight, particularly poor and working-class youth of color who do not directly benefit the neoliberal marketplace. The prominence of militarized discipline as a strength of the MEI is particularly interesting as the neoliberal rhetoric of discipline has worked its way into the decisions of parents, the behavior of cadets, and everyday practices of teachers at the MEI. That is, parents and teachers at the MEI view the neoliberal rhetoric of discipline as a solution to an underfunded and inequitable education system rather than alternative solutions such as equality of educational resources or funding.

Uniforms

The uniform is another aspect of the MEI that attracts both parents and students. The MEI uniform, which is the same for both boys and girls, initially consisted of dark blue slacks, a light blue button-down shirt with the school mascot embroidered on the shoulder. However, this was later changed to camouflage battle dress uniform (BDU), black lace-up combat boots, and a matching camouflage cap. The BDUs are made out of a thick durable material. These items are purchased at local military supply stores or through online military supply sites for approximately $75.00. The uniform is not only an affordable option for many working-class families; its wash-and-wear durability extends the life of the garments so that parents are less likely to buy replacement garments throughout the school year. I asked Commandant Wilson (who initiated the change when he took over the school in the third year) the reason behind the switch. He stated simply that switching the daily uniform from "dress blues" to BDUs was due to the durability and ease (no ironing) of the BDUs. While this may be partially true, the switch was also a powerful visual symbol of the militarization of the MEI. The dress blue uniform was in many ways unremarkable. Students could be easily mistaken for students from surrounding private schools. What the BDU accomplished was to mark the cadets visually in a militarized way. This sets MEI cadets apart from other students in the community and, as I argue below, facilitated the conferral of social status.

There are two main reasons why the MEI uniform is a point of attraction for parents and students. First, parents, students, and teachers see the uniform as alleviating peer pressure and material competitiveness among students. Students at the MEI are sensitive, like most teenagers, to being different and not fitting in with their peers. Mr Park sees the uniform as very important for working-class parents who send their kids to MEI, "Some parents and families can't afford to buy clothes so it becomes a socio-economic factor." Additionally, the MEI uniform also thwarts teasing based on material inequality. Angela, a White seventh grader, liked the uniforms because "if you are wearing the same thing, other people can't talk about you or your clothes or anything. So, you are all the same...one unit. It's cool." Lawrence, a Black eighth grader, also liked the uniforms for similar reasons:

> Cuz, at other schools you can't wear special uniforms where everybody wears the same. We all wear the same uniforms here. But at different schools we wear different clothes from other people. So they won't say that you [sic] clothes is ugly. And at this school they won't say that cuz they all have the same uniform...nobody gonna make fun of you.

Susan Uppal goes so far as to argue that uniforms are one of the most important factors of MEI's success:

> I think the number one component of the MEI that sets us apart is the uniforms because there is no peer pressure of who gets what clothes. I think when you take that factor out of the school it takes a lot of problems away with it. Because in a traditional school, if they don't have the right shoes, if they don't have the right pants, if they don't have the right clothes, then, they are less of a person. Well, here, they all look the same and it is based on their accomplishments and who they are, not what they wear.

Thus, the MEI uniform alleviates material peer pressure among the students. Several MEI teachers also argue that the uniform assuages gang influence. For instance, Mr Thomas argues that students at the MEI "don't have to worry about, you know, colors, the colors you see at other schools with the gang affiliation and the hats and that type of crap. They don't have to deal with that because you are wearing a uniform. I think the parents appreciate that as well." Mr Park also agrees: "In the public school system kids can...the ones that don't have uniform codes can represent themselves whether its gangs [or] whether it's sex appeal." Thus, the MEI uniform is well received for alleviating not only gang influence but also material inequality and teasing.

The second reason for the strong support of the uniform by parents, staff, and students is that the uniform confers social status. The uniform is a symbol of status for cadets and represents the success of not only the school but of students enrolled as well. Many cadets brought up how the uniform makes them feel good about themselves and proud, especially when worn in public at parades and community service events. Angela, a Latina ninth grader, stated that the uniform "makes you look good and people look at you while you're good and everything. And, I think I wanna be like that. You feel better." Additionally, Erika, a Latina seventh grader, felt good about her uniform: "People realize who we are and what school we go to. How good we are." For these cadets, being part of the MEI and wearing the uniform confers social status and is a source of self-esteem. This is not surprising as the uniform is militarized. In militarized nations and cultures like the United States, the military is valued and held in high regard (Enloe 2000). Those things that take on values and characteristics of the military also take on a similar status and value in society; this is the case with the MEI school uniform. Not only is the school uniform completely militarized, it is a powerful visual demarcation that confers militarized status upon MEI students.

Additionally, for kids whose parents work in landscaping, nearby casinos, or auto body shops, the MEI uniform may be the nicest clothes they own and their only source of visible material status. Mr Thomas sees the status of the uniform as far more reaching than just the kids and their families but to the community of Eastmoore as well:

> The neighborhood here, the area here, the whole Eastmoore community is a working-class community. So they love to see their kids, I think most parents want to see their kids do better than they did. And so, I think that is one of the reasons why the parents send their kids here, because they have that opportunity to do better, you know. So, there is a huge status thing here with that because with the uniforms especially. First impressions are huge, okay, and if you have parents that are together, okay, and one parent has a kid at the MEI walking toward that parent group; of course they are going to smile and be proud because it is a status issue. It is almost like private school…almost. So, there is a huge status thing and the parents are proud that their kids are in this school and yeah, there is a huge status thing.

Ana Lopez, a parent, also loves the uniform: "It shows who we are, we're different." The militarized uniform separates the students at the MEI from other teens in the area. Mr Thomas adds that the uniforms are important for the parents beyond their affordability, but also because the uniforms confer status on their children: "They are thrilled because they see them in their uniforms which is huge. And they get a big, warm fuzzy because they are in the uniforms." Claudia states that "they make fun of us or whatever, but it is?... I mean we stick out from any of the high schools."

Indeed, as Mr Thomas stated, many students at the MEI are under the impression that it is indeed a private school. They based this assumption on the admission and uniform requirements. A particularly outspoken eighth grader, Bryce, was confused when I explained to him that the MEI, although a charter school, is a public school: "But isn't this school private? Because you have to pay for the uniform? That's what everyone says." The confusion over the fact that the MEI is not private but a public school illustrates, I think, the racialized discourse of neoliberalism and privatization in which private organizations and schools are linked with whiteness and quality while public schools are connected to brown and Black racial groups and low quality (Haymes 1995). Even though the MEI is not a school with a majority White enrollment, the students and the community perceive the MEI as superior over other schools for reasons detailed above (academics, discipline, uniform) and thus assume that the MEI is a private school.

In short, the MEI uniform is an integral part to the success of the MEI. Not only does the uniform mark the cadets who attend the school as different and better than those who attend the local public schools, it also alleviates teasing and peer pressure among cadets. Both these reasons are particularly important for working-class teens and their families as the uniform simultaneously downplays class inequality while conferring social status through militarization.

Conclusion

Charter schools are one of the most popular neoliberal educational reforms. According to neoliberal ideology, charter schools improve student outcomes through competition, choice and the discipline of the market (see chapter 1 for a full discussion of neoliberalism and education). The MEI is no different in this regard relying on neoliberal rationale— founded to provide parents and students a choice and relieve crowding in the underfunded, academically struggling, and violent Eastmoore Middle School. What is unique about the MEI is the militarized structure of the school. The Eastmoore School District could have opened any variety of schools, but they opted to open a militarized charter school.

The logic behind this decision illustrates the nexus of disciplinary neoliberal logic and militarization. The neoliberal state has dismantled and gutted the welfare state, replacing it with a punishing and surveillance state that controls marginalized and unruly populations through militarized discipline. As outlined in chapter 2, militarized public schools, and in this case the MEI, transform "dangerous," poor youth of color into obedient citizens. However, neoliberal discourse is not implemented solely from the top down, but is also worked into the practices and discourse of the parents, teachers, and students at the MEI as they act within inequitable social conditions and internalize the neoliberal responsibility of self discipline and control. Thus, parents and teachers "buy in" to the logic of neoliberal choice and militarized discipline as it offers agency and an alternative to the inequitable and violent school district. Parents consent, in Gramscian terms, to militarized education and *choose* to enroll their children in a militarized charter school. This is more indicative of an underfunded and inequitable education system that limits the choices and participation of marginalized groups rather than true empowerment and agency. Parents in the Eastmoore community are making a Sophie's choice between violent and poor performing schools and a militarized charter school with a spotty academic record. Due to the inequality of the education system as well as societal inequities across race and class, parents and students at the MEI are not able to exercise real choice or practice true agency or empowerment. The false promise of freedom of the neoliberal market through choice is only available for those who can afford it, for those who are privileged and positioned within the neoliberal economy to be consumers of education.

In part, the MEI is a successful militarized school because of the circumstances of the Eastmoore community and the life chances, raced, and working-class status of the families that enroll their children at the MEI. The MEI and the Eastmoore School District successfully took advantage, through neoliberal militarized logic, the structural gaps and inequalities in the Eastmoore School District, namely underfunded, overcrowded, and violent schools, to open a militarized charter school. The MEI, with limited space and funding, is then able to selectively enroll students based on previous academic performance, setting the school up for performance on state tests that is superior to surrounding schools, in particular Eastmoore Middle School. Thus, the racial tension, presence of drug and gang activity, and academically failing and overcrowded local schools push parents toward the MEI. These push factors coincide with complimentary factors that draw, or pull, parents and students to enroll at the MEI. Parents and teachers perceive the militarized discipline-based structure as important for MEI, which will keep students safe from violence, racial tension, and gang involvement. Meanwhile, the academic record (although contradictory) of the MEI is superior to surrounding schools, and the militarized uniforms confer social status within the working-class Eastmoore community, marking MEI cadets as different and superior. This combined with the fact that the MEI has limited student enrollment results in students, parents, and the local community perceiving the MEI as an elite institution similar to that of a private school. The fact that the alternative to failing local schools is a militarized charter school was not the primary factor in the cadets' and parents' decision to attend the MEI, but rather a consequence and outcome of social inequality.

4

Reading, Writing, Arithmetic, and War: Militarized Pedagogy and Militarized Futures

> I really don't like the military. I mainly went into the military because my mom wanted me to.
>
> —Monica, an eighth grader

Since 1973, the US military has relied on volunteers to fill its ranks and is the single largest employer of youth under the age of 25. The US military is intimately partnered with neoliberalism utilizing it to bolster US financial and corporate interests globally and to discipline and control unruly citizens marginalized by neoliberal practices. The involvement of the United States in the Global War of Terror, the extensive presence of the US military across the globe, and the ongoing militarization of society create a continual demand for recruitment and retention. The US military set aside $913.8 million dollars for recruitment in 2013 (Emmanuel 2012) and one of the primary recruitment targets is youth, easily reached through the militarized public education system.

In 2002, President Bush signed the NCLB Act. Embedded within this Act is a clause that mandates that public schools must allow military recruiters access to the private information of students. Schools that deny access risk losing valuable federal funding (Buck 2004). Parents may sign waivers to keep their child's information confidential, but few parents actually know about, or utilize, the opt-out waiver. Those parents that have filed waivers on behalf of their children found the process a harrowing journey of red tape and bureaucracy.

Following the enactment of NCLB, less than 1 percent of public schools in 2005 denied access to military recruiters compared to 8 to 10 percent in 2001 (Associated Press 2006). Additionally, as mentioned in chapter 2, militarized educational programs such as the JROTC and TTT have become increasingly popular within public schools and particularly with the rise of neoliberalism. Funding for JROTC programs, where almost half of participants join the armed forces, increased 57 percent between 2001 and 2006 (Nazario 2007). In 2011, 44 percent of youth aged 16 to 21 years said they had spoken to a military recruiter (Department of Defense 2011).

The US military's focus on the recruitment of youth violates the UN Optional Protocol on the Involvement of Children in Armed Conflict to the Convention on the Rights of the Child. The United Nations implemented the protocol in February 2002, and it was ratified in the US Senate in December 2002 and is the same year President Bush signed NCLB Act. The Operation Protocol sets the volunteer recruitment age at 17, makes it a violation of the international treaty to target youth under 17, and requires that "military recruitment activities directed at 17-year-olds be carried out with the consent of the child's parents or guardians" (ACLU 2008). The recruitment of children by the US military is also in opposition with the Child Soldiers Prevention Act passed by Congress in 2008, which prevents arms trade with countries that recruit child soldiers. However, the United States itself regularly targets youth younger than 17 through the NCLB and JROTC programs, Middle School Cadet Corps[1] (similar to JROTC), ASVAB[2] testing, the Delayed Entry Program,[3] and marketing strategies that focus on youth culture such as video games, films, and music (Chery 2003). The *School Recruiting Program Handbook* published by the US army, which is distributed to all recruiters, emphasizes recruiting teens below 17 years: "first to contact, first to contract... that doesn't mean seniors or grads... If you wait until they're seniors, it's probably too late" (2004, 3). Youth, then, are a primary target for US military recruitment efforts.

As I argued in chapter 2, poor and working-class youth of color are overrepresented in the bottom ranks of the military and are funneled toward military enlistment due to the violent outcomes of neoliberalism: inequitable education, racism, and poverty. As illustrated in chapter 3, the historical inequalities of the Eastmoore community and school district push students and parents to attend and enroll at the MEI, while the disciplinary structure is perceived by parents, teachers, and students as a way to keep Eastmoore youth safe from violence in the community and ensure a brighter educational future. The students who attend the MEI are overwhelmingly working-class youth of color and meet the military's prime demographic criteria for recruitment through the economic and political marginalization of neoliberalism. However, not all methods of recruitment are as overt as those discussed in chapter 2. Socializing youth from a young age into a culture of militarism fosters not only acceptance of the military and a militarized understanding of the social world; it increases the chances that youth will develop a militarized identity that is depoliticized and uncritical. Militarism is the handmaiden of militarization. This chapter argues that students at the MEI are socialized into a culture of militarism, which results in militarization of students and development of a militarized identity and understanding of the social world. This leads students to view the military as an attractive and equally beneficial career choice as higher education.

Militarized Pedagogy: Discursive Symbols and Language of the MEI

Neoliberalism is an inherently antidemocratic economic and cultural project that perpetuates antiintellectual and anticritical approaches and understandings of society. Neoliberalism is also a pedagogical project as it spreads the logic of the free market and shapes the way in which we, as subjects of the neoliberal order, understand and make sense of social inequality, social institutions, freedom, and democracy (Giroux 2004). Neoliberal educational reforms, particularly in terms of corporatization and militarization of public schools, actively work against critical pedagogy, a practice that provides students with emancipatory knowledge and experiences to challenge, understand, resist, and shape the social world and unjust social processes and relations. Schools, as sites for the potential development and dissemination of critical pedagogy, have a responsibility to provide students with knowledge and skills that empower them to influence and critically participate in a democratic society as critical citizens and to foster democratic subjectivities. Unfortunately, schools are institutions and social sites that more often than not legitimate and create social inequalities as well as perpetuate dominant ideologies. This is particularly true with the increase in neoliberal educational reforms since the 1980s. Pedagogies, then, cannot be examined outside of politics, and politics must be understood as inherently pedagogical.

As outlined in chapter 2, neoliberal educational reform coincided with and perpetuated militarization of public schools and educational reforms. The militarization of public schools results in a depoliticized and militarized pedagogy that is uncritical and undemocratic. Such a militarized pedagogy militarizes social relations, institutions, policies, attitudes, desires, and identities, and replaces critical pedagogies with uncritical and undemocratic militarized pedagogies. Militarization of everyday culture within schools is achieved on the back of militarism and is accomplished at the MEI through the role of everyday militarized meaning-making through signs, symbols, symbolic relationships, language, and material conditions.

Theorists of critical pedagogy examine how how everyday practices create and recreate social power as well as resistance and argue that one way in which schools perpetuate and create social inequalities is through discourse. For example, Apple (1990, 2013) utilized discourse as a tool to argue that the everyday practices of schools, such as curriculum choice and organization, produce "official knowledge" that privileges and legitimates dominant groups. Giroux's (2005) use of discourse indentifies unequal social power relations within schools that are a reflection of the broader power relations within society. He argues that schools, much like other cultural institutions, perpetuate uncritical acceptance of social power relations, and that education is fundamental to democracy, as a critical citizenry is needed for democratic participation and that schools can be a vehicle for transformative and critical citizenship. In one of his more recent works in critical pedagogy, Giroux (2011) illustrates how neoliberalism is detrimental to democracy and the ways in which neoliberal ideals have infiltrated education. He argues that the critical aspect of education is at risk of being lost and replaced with an uncritical discourse of individual freedom and market-based solutions that perpetuate inequitable education and societal relations. Finally, McLaren (1998, 1999) examines the everyday culture, practices, and identities of teachers and students and argues that classrooms are sites in which larger societal relations of power are played out as well as resisted.

Discourse is an analytical tool used to examine power relations in schools and is a theme lifted from Foucault (1972). Discourse is the ways of composing knowledge and the practices that shape relations of power and subjectivities. It is a system of meaning making that shapes the way we understand and define our social world and ourselves. Discourse outlines what can be said and thought, when such things can be said, and by whom. However, discourse is not only important in what is said, but also what is not said or what is excluded. Thus, inherent within discourses are power relations or who has the authority to speak and when and whose voices are heard and whose are silenced. It is within these alternative discourses that are marginalized and subjugated that resistance to hegemonic practices has the potential to emerge.

Discourse shapes not only social interactions and the production of knowledge but also institutions such as schools. For example, students at the MEI are referred to as "cadets" rather than students. This practice shapes the way youth at the MEI understand themselves, their identities, and even how they interact and understand their relationships with other students. Cadets at the MEI are militarized through this discursive practice of the school and connect themselves to military values and ways of understanding. This is further illustrated by the fact that the cadets at the MEI refer to each other and understand their position within the school in terms of military rank rather than grade level.

Discourse is an important component of the construction of subjectivities that produce not only social coercion but also the Gramscian notion of consent (Foucault 1972). The symbolic (language and rituals, for example) and material (curriculum, equitable funding, safety, cleanliness) conditions of a school result in the development of a particular way of understanding and interacting with the social world and cannot be removed from broader societal social contexts and structures. These forces, social context, and symbolic and material conditions of the school need to be viewed in a dialectical relationship that in turn shape the discourse and pedagogy of a particular context or school. This is particularly true within militarized schools such as the MEI. Pedagogies influence subjectivities, and militarized pedagogies mold students at the MEI into a militarized understanding of the social world as they learn to view their lives and opportunities through a militarized lens. In the following sections, I argue that the discursive uses of symbols (e.g., the school mascot, cannon, flag ceremony) and language (e.g., the school motto, hierarchical structure of the school, school fieldtrips) utilized at the MEI shape cadets' subjective understanding of the social world as a subtext of the culture of militarism at the MEI. This results in a militarized pedagogy at the MEI that perpetuates the rationality of militarization as well as the militarization of students, their identities and their futures.

Symbols of Military Culture

Symbols are an important component of discourse and meaning-making. Symbols construct subjectivities, but also shape social understanding, and in this case a militarized pedagogy links cadets at the MEI with a militarized worldview. The discursive symbols of the MEI include the school mascot—The Rough Riders, the cannon and flag ceremony, "Rough Rider Rations," and the MEI uniform. The MEI mascot is the "Rough Rider." The Rough Riders were a volunteer cavalry unit under Theodore Roosevelt that was influential in the battles of Las Guasimas and San Juan Hill in Cuba and the Battle of Kettle Hill. A man wearing a Western-style hat riding atop a speeding horse with a whip in one hand and a rifle in the other symbolizes the Rough Rider mascot. The logo is a take on the Winchester Rifle logo and is emblazoned on school letterhead, the student handbook, registration packets, student uniforms, and the yellow polo shirts worn by teachers, administrators, and other support staff. The role model for the school, and what the school is literally represented by, is a man, a volunteer solider who is regarded as a legend, a protector, and a hero. When asked what the logo meant, a 12-year-old male cadet, Michael, summed up the meaning of a Rough Rider in the following way: "Go to the army, you will save people and stuff. That's what a Rough Rider means to be." Thus, for many students at the MEI, the Rough Rider is viewed as a hero and a warrior. Connell argues that "the figure of the hero is central to the Western cultural imagery of the masculine...Armies have been freely drawn on this imagery for the purposes of recruitment" (1995, 213). Thus, many students at the MEI identify with the masculine imagery of the hero and protector and construct a militarized masculine identity around and through this (as well as resistance to this; this will be discussed in more detail in chapter 5). When asked why he plans on joining the military after high school, Michael replied, "[Because] I like saving people". Thus, a 12-year-old student knows that being a soldier entails being a hero and clearly sees himself as fulfilling this role.

The second discursive symbol of MEI's militarized culture is the school's cannon and flag ceremony. The cannon was built by Major West and uses propane to "fire." The cannon is fired every morning to recognize the raising of the California State and US flags. The firing of the cannon, while benign as it is just an ignition of propane, is loud and is heard across the small community of Eastmoore. The firing of the cannon takes place beside the flagpole, which is positioned directly outside the main building facing one of the biggest and busiest intersections in the city. The cannon firing ceremony consists of two teams of students chosen by the teaching staff. Each team is comprised of three students who are responsible for readying the cannon as well as raising and lowering the flags every morning and afternoon. Both students and teaching staff consider it a privilege to be chosen to participate in the cannon and flag ceremonies. Students—let us not forget they are indeed still children—see the ceremony as an extra privilege because they are excused from class ten minutes early to lower the flag at the end of the day. Major West actually fires the cannon, and when this occurs, the leader of the cannon team yells "Present Arms!" and the cannon team members salute the flags as they slowly work their way up the pole.[4]

The cannon and flag ceremony works as discursive symbols to socialize the cadets to be patriotic citizens through militarized rituals and pins patriotism to legitimate or state-sanctioned violence. The cannon and flag ceremony is not so different from the US military flag ceremonies and color guard regiments. Thus, through this militarized ritual, students at the MEI connect their everyday school processes with formal military ceremonies. The discursive ritual is a militarized pedagogy that equates educational values with military values and students with soldiers. In addition, the central location of the MEI within the town's center incorporates military culture into the community as the cannon is heard daily and the flag and cannon ceremonies are public displays of militarism.

A third discursive symbol of military culture at MEI is the Rough Rider uniform, which is the same for both boys and girls. The uniform initially consisted of dark blue slacks, a light blue button-down shirt with the school mascot embroidered on the shoulder. However, this was later changed to camouflage BDUs, black lace-up combat boots, and a camouflage cap. The initial uniform was then relegated to "dress blues" and worn during ceremonies or special events at the school. The BDUs are made out of thick durable material and are purchased at local military supply store or through online military supply sites for approximately $75.00 and are similar to the uniforms worn by active military personnel. Although the school's justification for the uniform change was lower cost and durability, the choice reflects the desire of the school administration to reinforce military subjectivities among its students as uniforms shape how identities are performed (Craik 2005). It is quite startling to see 300 young people dressed in the uniform of the largest national military. The scene at the school becomes not much different than the pictures and videos broadcast from the cities of Iraq and Afghanistan or what you would expect to see on any US military base. This is especially apparent during the lunch break when hundreds of camouflage-draped students spill into the main courtyard, making lines for the cafeteria and scurrying to save tables for friends. One of the older and physically largest boys at the school told me on several occasions that he is often stopped on the street or in stores when he is in uniform and asked about his military service. Thus, through the uniform these students literally embody soldiers so much so that when physical size and age approach standards for military enlistment, cadets at the MEI are mistaken in the community for actual enlisted soldiers.

The MEI uniform is also important because it not only exemplifies the discipline and attention to detail expected in the military, but it is also a uniform that is actually worn by US military personnel. Parents and cadets purchase the uniform at local military supply stores. The MEI uniform is also a status symbol not only to students but also to the parents and community of Eastmoore. As discussed in chapter 3, students feel proud of the attention conferred on the uniform by parents and the local community. Pride in the uniform is also part of the military and its status in the US culture. Cadets at the MEI identify with and through the symbol of their uniform.

Finally, one of the most interesting discursive symbols of military culture at the MEI is the Rough Rider Ration. The Rough Rider Ration is a homemade spin-off of the army's MRE (Meal Ready to Eat) and is given to each cadet during overnight fieldtrips. MREs are typically given to soldiers in combat or field situations in which typical food preparation and service is unavailable. The Rough Rider Ration is a sandwich-sized plastic bag that compactly holds one canned wet ration (e.g., Spagettio's, Ravioli, etc.), one small can of fruit cocktail, 2 instant cocoa mixes, 1 bag of chocolate M&Ms, 2 cookies, and 2 sticks of gum. Major West explains to the cadets that the Rough Rider Ration is as important as MREs are to military troops on the battlefield. The Rough Rider Rations are often given out to community and school board members as souvenirs. The Rough Rider Ration is one of the more interesting symbols at the MEI as it creates a strong bond between the cadets at the MEI and the ground troops of the US military in stimulating actual conditions that troops would experience during a tour of duty. The Rough Rider Ration literally draws parallels between student life and soldier life.

The MEI uniform, the school cannon, the Rough Rider mascot, and the Rough Rider Ration are discursive symbols of the MEI that militarize cadets' understanding of themselves and their social worlds. Although the cavalry and the cannon are no longer in formal use by the US military, the use of such militarized symbols at the MEI links the cadets directly to the US military, fostering a culture of militarism at the school, but more importantly militarizing the school, its pedagogy, and the cadets. The use of these militarized symbols creates an avenue for MEI students to identify directly with the military and military culture. This is no clearer than the fact that students at the MEI are referred to and refer to each other as "cadets."

The Language of Military Culture

Language is also an important discursive element of the militarized pedagogy of the MEI. A poignant example of the use of language is the MEI's motto: "Stay Alert! Stay Alive!" The origin and implementation of the phrase "Stay Alert! Stay Alive!" as the school's motto was not a planned aspect of the curriculum at the MEI, but rather a spontaneous event that fit the situation and "stuck." During the first few weeks of the opening of the MEI, when most cadets were coming in contact with military culture and etiquette for the first time, First Sergeant Strong used the motto as a reminder to the cadets to pay attention and to keep focused throughout the school day, especially during the classes and times that focus on the military aspects of learning, such as morning formation and military marching drills. The two phrases eventually came to be a standard expression of the learning environment at the MEI and were later officially coined as the school's motto by Major West. For example, at the end of morning formation, the cadets shout out the school motto in unison, and Mr Jones uses the motto in relation to the goal of overnight fieldtrips by stressing to cadets the connection between survival and battle and asks cadets to consider what it would be like if they were ever "posted out somewhere with the military." Harrison (2003) argues that this idea of "combat readiness" or "alertness" is imperative to the organization of the military: "to achieve and maintain combat readiness, the military must turn ordinary human beings into the kind of people who at any time can be mobilized to make war" (73). Thus, turning "ordinary human beings" or ordinary students into cadets "mobilized to make war" is done through equating the school motto with battle and survival. Many cadets make exactly the link between the motto "Stay Alert! Stay Alive!" and combat readiness. During the focus group interview, the cadets were asked what the school motto meant. One student excitedly related the motto to the battlefield. "It's like the battlefield...if you don't stay alert, you are going to die." In a one-on-one interview, Alberto, the Lieutenant Colonel during the first two years of the school (equivalent of student body president), drew similar comparisons: "Stay alert means do well...won't stay alive if not listening to instructors." A male seventh grader fervently stated, "'Stay Alert' means to keep your eyes peeled!" Angela, a seventh grader who currently lives in foster care, said the motto meant to "Watch out." When pressed for what she needed to watch out for, she replied "anything." These kids are on watch; they are ready for anything that might come their way. In a sense, these kids are "combat ready" and view themselves as cadets or soldiers, not as students.

The language used at the MEI reflects militarized pedagogy at the school. The very fact that cadets, as well as teachers, link the motto of the school to "survival" and "battle" exemplifies the permanence of military culture and a war readiness mentality. The choice of words for the school's motto is pregnant with military imagery. The first phrase—"Stay Alert!"—is actually benign and makes no direct link to battle or militarism and in combination with the second phrase does not contain connotations of battle or survival. Consider some options that could have been chosen for a motto for a public school: "Stay Alert! Learn!" or "Stay Alert! Study Hard!" or "Stay Alert! Stay Focused!" However, by equating "alert" with "alive" in the school's motto, the focus of the school turns away from learning, fun, and growth to physical survival and violence.

The hierarchical structure of the school is also an important discursive element of the militarized pedagogy and construction of militarized identity at the MEI. The school utilizes the US Army ranking system. Cadets are assigned the lowest rank of "private" during their first year at the school and are able to move up the ranks by passing certain exams that cover military knowledge and by being nominated by instructors at the school. The student body president, for example, is referred to as the Lieutenant Colonel. Cadets at the MEI speak of themselves or other students not in terms of grade level but in terms of military rank. Cadets refer to other cadets as privates and sergeants rather than seventh or eighth graders. This hierarchical system even applies to the principal, the commandant, and any staff or teachers with military experience, such as First Sergeant Strong. Thus, the discursive practice of speaking to each other in terms of military rank furthers the construction of a militarized identity as students view themselves and others as fully ranked members of the military.

Finally, the themes of the four official fieldtrips or "operations" also exemplify militarized language at the school. The themes of the fieldtrips are suggested and voted on by the cadets. Examples include "Operation Fire Ant," "Operation Fire Bee," and "Operation Fire Cat" and finally "Operation Fire Dragon," a BBQ party with an awards and decorations ceremony. Militarized language inherent in the themes of the three fieldtrips and the final awards and decorations ceremony refers to battles or "operations." The latter term conforms to catchy titles given during the past 15 years to wars and battles by the US military, such as "Operation Desert Storm" and "Operation Iraqi Freedom." As a result, the "operations" link cadets with battle and survival. It is also made for parents and guardians, thus building a bridge for militarized language and symbols between school and home. Thus, the militarized pedagogy of the MEI militarizes not only the students but also their families and their homes.

Teaching the cadets basic survival skills is the stated goal of the overnight camping trips. A backpack list contained in each information packet details the camping or hiking items the cadets need to bring with them, such as socks, extra shoes, first aid items, toilet paper, metal fork and spoon, two liters of water, potato, carrot, and onion. The cadets must also bring a knife, which will be used "to cut up their food" and must be a "hunting, lock-blade, [or] buck" style knife. This list contains 16 different articles, five of which are specifically labeled as "survival"; that is, the word "survival" is in parentheses beside each of the five items. These survival items are a lighter, a whistle, a candle, a mirror, and twine or long shoelaces. Only two of the five "survival" items have a specific purpose on the overnight trip. The lighters are used to start and maintain campfires.

The mirrors are used for signaling by reflecting sunlight as part of an exercise where the cadets must locate a volunteer chaperone hiding in the surrounding hills with an ice chest full of chicken (that night's dinner).

The theme of survival is prevalent in the fieldtrip information packs with the items that cadets need to bring with them on the overnight trip, specifically those labeled as "survival items." The cadets placed much emphasis on the five "survival items." One cadet stressed the importance of the lighter in this way: "in case you can't keep the fire started. There's no lights out there, so in case you run out, that's your only light." To listen to this cadet stress the importance of the lighter to "keep the fire started" and as "your only light," the student feels he or she is on a solo adventure rather than taking part in a well-organized, highly supervised school function. Cadets also emphasized the style of knife required for overnight trips. Many of the cadets in the focus group stated that "carrying knives" was one of the best parts of the camping trip. As mentioned earlier, the style of knife must be one for hunting, lock-blade or buck (there is no blade length limitation). It is clear that what is required is a knife that can be either folded or sheathed for somewhat ironic safety reasons. However, as the knife is only to be used to prepare food, a simple paring knife with a plastic cover would appear to be sufficient. However, by specifying that the knife must be a "hunting, lock-blade [or] buck," the knife and the act of carrying it are explicitly linked to the theme of survival, a sort of proto-battle against the wild.

For several of the male cadets in the focus group, the knife was a status symbol, and they boasted, "no one else gets to do it" (no other seventh grade students are allowed to carry knives at a school function). It is doubtful whether students would view a simple kitchen or paring knife in a similar fashion. Being a "survival" item, the students accord it special status. One particularly boisterous male cadet, Michael, has had his knife taken away on all four camping trips. Michael was quite proud of this and recounted why he had his knife taken away on the latest trip. "Yeah, I had my knife and I would like to stab it into this bench because it was all crappy. I was like (makes the motion of repeatedly stabbing the table)." Thus, for these students the knife is not a tool with utility (i.e., preparation of food) but rather an item *needed* for survival, and because of that an item with status. Yet, they downplayed the importance of bringing their own food, water, warm clothes, or tent items required on the fieldtrip. Items that are actually needed for human *survival*. The cadets internalized the survival theme and viewed the overnight excursion as a test of their survival skills rather than a school-related educational activity. As a result, the theme of "survival" is linked directly to battle and war, and not necessarily to the actual needs that are basic for human existence.

Conflating Military and College

Through the use of discursive militarized language and symbols, cadets at the MEI are awash in military culture through which they become militarized subjects with militarized subjectivities. Through the militarized pedagogy of the MEI, cadets themselves become militarized and understand education to be synonymous with the military and military service. For example, Monica, an eighth grade Latina who is not particularly interested in the militarized aspects of the MEI such as participating in the drill team or flag and cannon ceremonies, has internalized the militarized identity and perceives herself as already part of the military: "I really don't like the military. I mainly went into the military because my mom wanted me to." Jack, a quiet a White seventh grader, also conflated the MEI with the military: "I thought it was a good school and that you get to have free college. That's what they said." Obviously, Jack is confusing the military's college tuition programs with the MEI thinking that enrollment at the MEI, much like enrollment in the armed forces, would result in paid college tuition.

Additionally, many cadets believe that the military and college are equally beneficial with equally beneficial results. Many cadets believe the military offers similar job and education opportunities as a college education. For example, Jason, a White eighth grader, says he is thinking of joining the military or going to college, but thinks joining the military "will be interesting and when you get out, you get better jobs and more money and stuff. I want a car." It is interesting that Jason did not talk about college being an avenue for a "better job" or "more money." Additionally, Miguel, a first-generation Latino, also considered the military to be an educational institution: "Well, they help you when you like, if you quit college you can go there. And they give you like your education. So you can practice both things." Thus, many students at the MEI are so immersed in military culture, so fully militarized that they not only see themselves as already part of the US military but also understand the military and education as similar institutions that help them to achieve similar goals.

Conclusion

As the lunchtime bell rings out across the central courtyard, the once-quiet courtyard is bombarded with commotion and noise as classroom doors swing open spilling forth green, uniformed bodies, half-zipped backpacks and the heavy slap of black combat boots onto the central courtyard. Watching the cadets at the MEI jostling for a place in the cafeteria line, rushing to save faded fiberglass lunch tables, and shouting to friends and classmates makes it easy to forget that these cadets are nothing more than students on the brink of summer break. It is easy to forget that these students are cadets studying military strategy, learning how to march, to salute, and equating educational lessons with survival. The MEI is about more than the simple education of its cadets; it is also the site for militarized pedagogy and the construction of militarized identities of some of the most vulnerable citizens in the United States.

Since 1973, the US military has relied on a volunteer force to fill its ranks and has targeted youth and primarily disadvantaged youth of color for military recruitment even though this stands starkly against two agreements signed by the US government: the UN Rights of the Child and the Child Soldiers Prevention Act. The US military utilizes schools as a sort of proto-weapon for the recruitment of youth emboldened through neoliberal structural constraints and educational reforms that limit the options for disadvantaged youth. For example, the army's recruiting handbook encourages recruiters to reach youth first (as opposed to colleges and universities), and the NCLB passes on the contact information to these very same recruiters. Militarized education programs such as the JROTC, TTT program, and fully militarized schools like the MEI work as militarized pedagogies spreading the logic and values of the military and military service.

For the cadets enrolled at the MEI, life becomes a battle not of educational skills or knowledge, but of survival. The discursive language and symbols at the MEI socialize cadets to understand the world around them in militarized terms. The use of militarized language in school practices at the MEI, such as fieldtrip "operations" and the school motto, militarizes students to view themselves in line with the military and as student soldiers on a metaphorical battleground. This becomes particularly powerful when combined with the material aspects of military culture such as the Rough Rider school mascot, the cannon and flag ceremony, the Rough Rider Ration, and the uniform. What is more, the militarized pedagogy of the MEI spreads militarism throughout the local community through the visual presence of youth in military uniforms and the daily cannon and flag ceremony as well as building a militarized bridge between school and home with the school fieldtrips or "operatioins."

The militarized pedagogy of the MEI militarizes students and allows cadets to conflate education with the military, but also uncritically accept the idea that enlistment in the armed forces is an equally beneficial choice as going to college. Since 2001, public schools have become successful avenues for military recruiters struggling to fill the ranks of the US Armed

Forces. The socialization of cadets at the MEI then militarizes cadets to view the military as necessary and beneficial to their life chances. As cadets at the MEI already view themselves as part of the military, this arguably increases the chances that MEI students will be open to military recruitment. Even though the MEI is fully militarized and utilizes a militarized pedagogy to disseminate a militarized understanding of education and the cadets' social world, not all cadets willingly consent or accept militarization. Cadets also actively resist such militarized structures and social relationships through and with their raced and gendered bodies as well as sexual practices and identities. Resistance to militarized social relationships and structures of the MEI is discussed in the following two chapters.

5

A Few Good Boys: Masculinity at the MEI

> If there is a guy caught out there and [a] machine gun is raking him. You [would be] running out there and grabbin' him and running back.
>
> —Than, a ninth grader

The militarization of US public schools, although intensifying since the events of September 11, did not start at this time but is part of the militarization of broader civil society and the "enforcement of global corporate imperatives" (Saltman 2003, 1). School militarization needs to be understood as a phenomenon that is linked to corporate globalization, which has steadily increased since 1945 (Amin 2004). In chapter 4, I argued that the militarized symbols and language at the MEI shape the identities and subjective understandings of the cadets, resulting in students viewing military enlistment and the pursuit of higher education as equally valid pathways. Recent research explored the militarization of public schools (Berlowitz and Long 2003; Giroux 2009; Robbins 2008; Saltman and Gabbard 2011), but few studies examine how militarization and militarism have an impact on the construction of gender and, particularly, masculinity. This chapter examines the construction of a hegemonic masculinity at the MEI, the nuances and effects of hegemonic masculinity for both boys and girls, as well as how militarized masculinity is resisted. This chapter begins with an examination of the literature on gender, militarization, and education. Connell's theory of hegemonic masculinity is utilized to understand the intricacies of the construction of masculinity at the school for both boys and girls. Ultimately, this chapter argues that hegemonic masculinity is exemplified at the school through the condonement of violence and the warrior hero archetype. Not all cadets at the school have access to or can capitalize upon the advantages of this particular form of hegemonic masculinity, specifically girls and Black boys. However, militarized masculinity is challenged and resisted by normatively feminine girls at the school by the conscious decision to make mistakes during drill, slowdown of physical training, and uniform violations.

Gender and Education

Gender is constituted through ongoing daily interaction (West and Zimmerman 1987). However, our daily interactions and actions are situated within larger social patterns and institutions. A large body of work exists that examines how elementary and secondary schools serve as social contexts that construct and maintain both gender and gendered identities (Eder 2003; Kessler et al. 1985; Thorne 1993). Much of this work examines the gendered differences between boys and girls and how schools perpetuate these differences. This is what Best (1983) refers to as the "second curriculum" or Kessler et al. (1985) refers to as the "unofficial school." Studies have examined differences in friendship patterns (Eder and Hallinan 1978), extracurricular involvement (Eder and Parker 1987), conversational patterns and rules (Gilligan 1982), as well as popularity construction (Adler et al. 1992). These earlier studies examined gender in terms of sex roles and viewed sex roles as oppositional and complimentary. This changed in 1992 when the report of American Association of University Women (AAUW 1992), *How Schools Shortchange Girls*, revealed that K-12 girls receive a lesser education than boys. The AAUW report turned the focus of research toward girls' educational disadvantages, resulting in a plethora of publication and scholarly work (e.g., Orenstein 1994; Pipher 1994; Sadker and Sadker 1994).

The focus on girls faded in the mid-1990s as the attention and research underwent a "boy turn" (Weaver-Hightower 2003). The turn toward boys focused on the troubles of boys and produced a laundry list of problems such as lower standardized test scores in reading and writing, more likely to be diagnosed with learning disabilities, behavioral problems, and expulsion (Kantrowitz and Kalb 1998; Kleinfield 1999; Knickerbocker 1999). The "boy turn" produced a large amount of research that argued that boys were in "trouble" and that schools were failing them. As Kimmel (1999) points out, feminists and the earlier focus on girls were blamed for much of boys' problems.

Although the way in which the boy turn blamed feminism for boys' troubles was problematic, it was successful in that it prompted research into the relationship between masculinity and education and the consequences this has for boys (Connell 1996; Mac an Ghaill 1996). For example, Francis and Skelton (2001) illustrated how male teachers use discourses of misogyny and homophobia to construct masculinity in classrooms, and Parker (1996) analyzes how physical education constructs masculinity for boys in public schools. While much work during the later part of the late twentieth century pitted girls against boys, there was a shift away from this type of analysis as scholars began to recognize that gender categories were not homogenous and there are just as many differences within gender as there are between them (Griffin and Lees 1997; Thorne 1993).

Militarization and Education

Schools are not only sites for the production of masculinity but are increasingly sites of militarization. As outlined in chapter 1, the process of militarization can be applied to institutions and organizations or even people and specific groups of people. The militarization of public schools ranges from teaching students to stand in single-file lines and follow orders, to school uniforms, to the implementation of programs such as the JROTC, TTT, and even militarized curriculum and fully militarized public schools. "The militarization of public schools is both a material and cultural project [and] can be understood as part of the rise of a 'warlike mentality' in the United States" (Saltman 2000, 84). This militarization of US public schools raises interesting questions regarding gender and, in particular, the production of masculinity. Militarization and education is an area of research that has been understudied, and it is to this endeavor that this chapter now turns.

The military is one of the primary sources of how masculinity is viewed in society (Connell 1992; Morgan 1994). Paul Higate (2003, xiii) argues that even "mundane military processes, such as military training, have major impacts upon individuals and groups." Military organizations provide "social and psychological resources for the reproduction and changing of individual psychologies." Thus, if simple processes of militarization affect and change the group, as well as the individual level, how does a "warlike mentality," militarism, and militarization within a public school affect what is learned and how it is taught? Does militarization within schools affect boys and girls similarly? How does militarization at the MEI influence the production of masculinity within the halls of the MEI? This chapter examines how militarized masculinity is played out and bound together, for both boys and girls, in the social space of a militarized charter school. Specifically, I argue that the construction of hegemonic masculinity at the MEI is bounded by race, gender, and sexuality as only White heteronormative males are allowed to capitalize on this particular form of hegemonic masculinity. Girls and Black boys who attempt to access hegemonic masculinity are particularly severely sanctioned. Not all cadets want to enact hegemonic masculinity as illustrated by feminine girls at the school who actively resist masculinity.

Masculinity, Militarization, and the MEI

Connell's (1995) "hegemonic masculinity" is one of the most influential constructs in theorizing masculinity. Rather than arguing there is a dominant or main role for masculinity, Connell (1995, 76) argues that hegemonic masculinity is "the masculinity that occupies the hegemonic position in a given pattern of gender relations" and is multiple, fluid, and always contested. Connell's theory outlines a hierarchy of masculinity within which men, and boys for that matter, are caught. This contested hierarchy of power includes complicit masculinities—men who do not adhere to all the characteristics of hegemonic masculinity but nonetheless benefit from it; subordinated masculinities—men who are suppressed by the definition of hegemonic masculinity (mostly gay men); and finally marginal masculinities—men who have gender power but lack power in other areas such as race, ethnicity, and class (Connell 1995, 80–81).

Connell's framework serves as a useful theoretical lens in examining particular gendered discourses and practices, such as the construction of masculinity at the MEI. However, it should be noted that it is often forgotten that Connell's theory of masculinity is fluid and nestled within particular social contexts and patterns of power. Connell's work is often used to construct types of masculinities (Miller 2001, 49–50; Pascoe 2007, 8) that read like laundry lists of characteristics of men rather than as a useful tool to examine the discourses and practices of masculinity. Additionally, much of the work on masculinity links masculinity with male bodies and with what men and boys do. This sets masculinity within a framework that essentializes the differences between males and females and fails to recognize the vast amount of differences within genders. This dichotomy rests upon categories that are assumed discrete, but are in fact socially constructed (Fausto-Sterling 1995), and ignores the possibility that masculinity can lay outside biology and the male body. It ignores the possibility that women and girls can enact masculine identities and practices (Halberstam 1998; Paechter 2006). Thus, utilizing Connell's model of hegemonic masculinity, I am not constructing an ideal type of militarized masculinity but rather examining how hegemonic masculinity and its inherent power relations are constructed at the MEI for both boys and girls.

In *War and Gender*, Goldstein (2001) gives a historical account of military values and masculinity and argues that the relationship between masculinity and militarism is not new. "Being a warrior is a central component of manhood, forged by male initiation rituals worldwide...Common features of 'warrior values' across cultures and time periods are closely linked with concepts of masculinity" (2001, 266–267). Higate and Hopton also see a historical relationship between militarism and masculinity but argue that it is reciprocal. "On the one hand, politicians have utilized ideologies of idealized masculinity that valorize the notion of strong active males collectively risking their personal safety for the greater good of the wider community...On the other hand, militarism and militarization feed into ideologies of masculinity through the eroticization of stoicism, risk-taking, and even lethal violence" (2004, 434). Several different research studies have explored how the military is essential for the construction of particular and context-bound hegemonic masculinities (Hicklin 1995; Jennings and Weale 1996; Woodward 2000) as "militarist values continue to have disproportionate influence on the ways in which hegemonic masculinity[ies] are both created and reproduced" (Higate and Hopton 2004, 444). As Morgan (1994) argues, "war and the military represent one of the major sites where direct links between hegemonic masculinities and men's bodies are forged" (168).

Thus, there are strong connections between masculinity and the military; the military and masculinity define one another and are bound up in a symbiotic relationship. Thus, military values and ideals influence the ways masculinity is produced and "is the major means by which the values and beliefs associated with ideologies of hegemonic masculinity are eroticized and institutionalized" (Higate and Hopton 2004, 436). Thus, the construction of US hegemonic masculinity is militarized, and the military is in the business of making men. The following sections will outline the characteristics of hegemonic militarized masculinity at the MEI, namely the construction of the warrior hero, the use of and glorification of violence, and how these characteristics influence and shape the gendered practices of both boys and girls.

The Warrior Hero

One of the main characteristics of US hegemonic masculinity is the idea of the protector, exemplified in militarized masculine identity as the warrior hero (Dawson 1994; Goldstein 2001; Woodward 2000). The warrior hero is strong, independent, masculine, male, White, and willing to risk his life for others. The warrior hero is a gendered archetypal who has traditionally reinforced patriarchy as the act of protecting is equated with men and masculinity, while the act of being protected is most often equated with women and femininity (Howard and Prividera 2004; Stiehm 1982; Woodward 2000). This stems from the belief that women are weaker than men, in need of protection, and from the general inequality between the sexes in regard to power, whether political, economic, or physical. This link between protector and protected is used to motivate not only soldiers in time of war but nations as well. However, the cadets at the MEI did not construct the warrior hero in this traditional manner.

Many cadets, both boys and girls, spoke about the warrior hero identity when asked what it meant to be a man in the military. Monica, a Latina seventh grader who spends much of her time discussing fashion, boys, and cheerleading, viewed men in the military thus: "they [men in the military] learn responsibility [and]...not [to] just think about yourself." John, a White ninth grader, thinks that "to be a man in the military is...to have the courage to be able to do something for someone else, like risk your life for their life." Similarly, Than, a Vietnamese ninth grader, set up the following scenario to explain militarized masculinity: "If there is a guy caught out there and [a] machine gun is raking him. You [would be] running out there and grabbin' him and running back." Thus, cadets at the MEI view men in the military as those who are willing to risk their lives for others. However, when cadets speak about this "other," it was not necessarily in gendered or sexed terms. This other, that is being protected or rescued, could be male or female, man or woman. Only Than referred to the sex of the person being saved as male or, in his words, a "guy."

As discussed earlier, the act of protecting is equated with men and masculinity, while women and femininity are in need of protection. Both girls and boys spoke of wanting to fulfill the role of the warrior hero and to risk their lives in the service of others. For example, Maricela, a Latina seventh grader who is active in military drill and spends much of the lunch break chasing and "beating up boys" with her best friend, wanted to join the military because she liked "how the military and how people risk their lives and everything and I wanna do that." Although she quickly added "but I am not so sure my mom is okay with that."

Interestingly, only one male cadet who I spoke with, Doug, ever identified the warrior hero identity as specifically involving the protection of women. This White eighth grader was one of the most involved cadets with the school and the military (he attends several military summer camps, reads military manuals in his free time, and intends on becoming a career marine). He described the warrior hero identity in this way:

Being called a man would be like putting his life on the line for any, for his country or anything. Um...like...for instance that bombing on that one airbase...Like if he had a girlfriend on the base, and the bombing was happening and shrapnel or what they call these...stuff and parts coming off the bombs. He would probably jump and at least cover the girl so she won't get hurt or anything.

Doug's answer was not typical of cadets at the MEI, and interestingly, he referred to the protection of military women, not necessarily civilian women. Unlike Doug, the majority of cadets at the MEI did not necessarily view the warrior hero in the gendered and sexed terms of protector as masculine/male, and those in need of protection as feminine/female. The cadets viewed soldiers as fighting for their country, helping people, and risking their lives for others, not necessarily women alone. So, if the warrior hero is not necessarily employed in the protection of women, how do the cadets at the MEI view women in the military?

Women and the Warrior Hero

The fact that most cadets did not construct the warrior hero in gendered or sexed terms in relation to the protector/protectee dichotomy could be attributed to the fact that cadets at the MEI view women as important additions to the military. Cadets at the MEI see women as able to fulfill the warrior hero image. For example, Manuel, a quiet eighth grader, argued that women in the military help the men out. "Well, the girls like...help them...the girls like have a different mind than boys and they...are smarter or something." A rather poignant example took place one particularly sunny afternoon as I spent the lunch period chatting with a group of six ninth-grade boys. This particular group of boys was always together, was highly involved in the school's military activities such as drill and flag, and was constantly discussing weapons, war strategies, and sex. Sarah, the school aide, described this groups of boys to me when I first started volunteering at the school with a sigh and a roll of the eyes saying "all those boys talk about is blowing stuff up and sex." Thus, when this particularly infamous group of boys was asked what they thought about women in the military, I was somewhat surprised.

Initially, the group of boys made jokes, laughing that women in the military gave male soldiers "incentive" and "something to look up to." The tone of the conversation changed, however, when Than, a mature ninth grader, stated that including women in the military was an issue of rights. "It gives more people rights you know that? And that works...back then it was only men that joined the military. It would make sense if women joined the military." At this point, to my utter surprise, other boys in the group chimed in and agreed with Than with comments like: "I believe women can be good in the military too" and "Yeah, if they set their mind to it." It was even eventually concluded among the group that women could actually excel in certain areas and outperform male soldiers. One of the group members, John, told me, "They [women] can be better in some areas. Like they can fit in small spaces and they're smaller and they can probably do maybe a little bit more because they are more agile and stuff." However, after much debate it was decided that although women had a lot to offer the military and could even outperform men in certain areas, women would definitely have to carry "smaller guns" because of their lack of physical strength.

A conversation I had with Doug traveled along a similar logic. He claimed that women should be allowed in the military, and that women offered unique skills to the military situations.

> Actually, I do because...most people would say that they don't but I do because if you think of it, without women in the military a lot of the branches would go down...Because now, the women are seeing how it is...if they're working at an ordinary place they're not doin' something for their country. So, if they want to do it they should have every right to go into the military.
>
> B.J.: Are there any more advantages to having women in the military?
> Doug: Well, if you're using that girl or woman as like an army spy or something. The women can actually get a ton more information than the guys can get. They have mind control over the men. So that basically whatever they ask they get. If they want information or something, they can easily get it.

Although most of the comments that boys at the MEI made about women in the military espoused equality, these statements smacked of sexism. Statements such as, "if they put their mind to it," meaning that if women tried hard enough they could equal men, or the fact that women would have to carry smaller guns because of their inherent physical weakness relegated women in the military to lesser status even though the boys viewed women as important additions to the military generally.

In conclusion, cadets at the MEI view women as equal members of the military, albeit in different terms. However, this equality is in regard to the opportunity to join and not necessarily in regard to equal treatment or equal performance capacity. The boys at the MEI saw women as offering a variety of skills that men do not possess or that men do not excel in, such as agility and espionage. However, these "advantages" are based on gendered assumptions. For instance, women in the military are more agile because of their smaller and weaker bodies and women are good at espionage because they can use their sexuality to win control over men's bodies and thus their minds. However, it is important to keep in mind that cadets at the MEI saw both boys and girls as exemplifying the warrior hero, risking their lives, and serving their country. While the cadets did not utilize the traditional gendered dichotomy to speak about the warrior hero, gender was still utilized as an organizing principle. Thus, the warrior hero identity at the MEI is applied to both girls and boys but is separated from the traditional assumption that the protector is male/masculine and those in need of protection are female/feminine. This contradicts previous literature on masculinity and the military that argues that militarized masculinity is strengthened by the gendered protector-protected dichotomy (Enloe 1989; Nantais and Lee 1999; Turpin 1998).

Violence

The use of and threat of violence are also an integral part of hegemonic masculinity at the MEI, as well as US militarized masculinity generally (Connell 1995; Morgan 1994). Although many parents send their children to the MEI in an effort to avoid the gang and drug violence of the middle and high schools in the area (as outlined in chapter 3), violence is prevalent at the MEI and is part of the school's culture and militarized masculinity. The culture of violence at the MEI can be broken down into two parts: the inculcation of weapon knowledge and the condonement of acts of actual violence. While these two parts are both masculine and militarized, only actual acts of violence are linked to hegemonic masculinity at the school, and it is only these acts of violence that are bound by race and gender.

Inculcation of Weapon Knowledge

The MEI is a militarized school, and the military is an implicitly violent social institution. The MEI has, in the words of Enloe (2000), come "to imagine military needs and militaristic presumptions to be not only valuable but also normal" (3). Thus, there is an implicit culture of violence at the school that is accepted and normalized. Part of this implicit culture of violence is the knowledge of weaponry. During the tenure of Major West, the school took several overnight fieldtrips during each school year. On these fieldtrips, MEI cadets learned to read maps and compasses, communicate with mirrors, start and maintain fires, and other "survival" skills. Additionally, each cadet got the opportunity to fire a rifle at a paper target during the fieldtrip. This activity was heavily regulated as Major West himself helped each student fire the gun, supporting some while they fired and giving gun safety advice throughout the exercise. Firing the rifle was one of the most popular activities during the fieldtrips, and cadets excitedly described the experience, quickly identifying who were the best shooters. Additionally, both staff and peers praised students who were good shots. One of the best shots, Marisol, informed me that she was going to be even better on the next fieldtrip as she was practicing her aim with her father's paintball gun.

What is interesting about this activity is that, first, all cadets participated. The firing of rifles was not reserved for upper grade levels, the advanced, or for any particular race or sex group. Second I did not hear any cadet state they did not enjoy the activity. Some students, mainly female, stated that discharge of the gun "startled them" or "was loud," but I never encountered a student who did not enjoy the activity. What this example illustrates is how the knowledge of weapons is a valued part of the curriculum at the MEI for all cadets. It is part of the general militarized culture of the MEI.

Another example of the instillation of knowledge of weaponry at the MEI is Shawn. Shawn is a Black seventh grader who has yet to reach puberty and remains the smallest and most child-like cadet at the MEI. He stands out at the MEI precisely because of his small stature. I have never actually seen Shawn when he is not singing, dancing, or nonstop talking. Shawn physically vibrates with excessive energy, and this often lands him in detention or the commandant's office. Shawn is well liked at the MEI, but most cadets treat him like a younger sibling rather than a peer. Girls constantly pat Shawn, and the boys either goodheartedly tease him or simply ignore him. Shawn is charismatic and takes it upon himself to inform all the adults at the school, on a continual basis, that he is "adorable" and that he and they "know it too." When Shawn is not plying for the attention of those around him, he makes realistic guns out of masking tape. The work and ingenuity that goes into making these guns is quite impressive. They are extremely realistic, true-to-size, and come complete with reloadable magazine clips and parts that can be locked and cocked. The guns, however, do not fire and are physically harmless.

Shawn regularly points his guns at his peers while emulating the discharge of his particular weapon. His favorite time to do this is during the lineup that occurs before each class. During the lineup, cadets are to stand at attention in single-file lines outside of each classroom. Cadets are reprimanded for moving, talking, or incorrect body position. This is when Shawn hides behind a larger cadet and moves his gun back-and-forth, mowing down the parallel lines of cadets. While the instructors at the MEI knew about Shawn's masking tape gun creations, Shawn was very careful to never point or "shoot" his gun at other cadets in situations where the staff, the commandant, or any other instructor could see him. He always "used" his gun surreptitiously. During the time I had spent at the school, I never once saw Shawn reprimanded for making or having the guns.

In a time of heightened concern over school safety, along with the growing list of campus killings with guns, it is almost implausible that the guns would not be questioned. However, if this behavior is examined through the lens of a culture of weapon knowledge at the MEI, Shawn's behavior is acceptable. Obvious knowledge of the workings of firearms is an accepted and normalized part of school culture and is also part of the underlying culture of violence at the MEI. The inculcation of weapon knowledge is pressed upon all cadets at the school, regardless of race and gender. This, however, is very different from acts of actual violence.

Acts of Violence

The second part of the culture of violence at the MEI is the condonement of physical acts of violence. One of the most poignant examples was an eight grade group science assignment. Cadets were formed into groups and told to use materials found around the house in which to conduct science projects. Three boys, Doug (mentioned earlier), Peter (one of the largest and most aggressive students at the school and White), and Hector (a light-skinned, amiable Latino, with a round face and freckles who played a very peripheral, supporting role in the science project as he never handled or fired any of the rockets), formed a group and made "rockets" out of two-liter plastic bottles. The plastic bottles were painted black and completed with halved CDs that served as wings. The group conducted their science project behind the school in the open soccer field. The group launched three different rockets. The first rocket rested on the grass and was shot into the sky using a combination of compressed air and water. The second rocket was fired by placing the butt of the rocket into a band of thick black rubber that was firmly attached to the metal gateway that opened onto the soccer field. This served as a type of slingshot that flung the rocket quite unsuccessfully into the grass. The third rocket was fitted into a large plastic pipe that rested upon Peter's right shoulder. Peter, dressed in full camouflage, knelt on one knee and squinted, while the other two cadets attached the hose of the air compressor to the back of the pipe. The "rocket" literally exploded out of the pipe with a loud bang and traveled several hundred meters into the soccer field. This was met with the cheers of classmates, the teacher, and several of the mothers who had come to watch the science project presentations.

At its heart, this science project was a great way to encourage cadets to be inquisitive about the world around them and understand that science can occur in everyday contexts with everyday tools. The science project was a good introduction to understanding the basic principles of physics and aerodynamics, but what is important is that these basic principles and ideas were explored and presented within a context of militarized violence. The boys' insistence on referring to the plastic bottles as "rockets" and the fact that the launching devices the boys chose to use and construct emulated realistic warfare devices illustrate how the content of science classes were militarized. Additionally, Doug, Peter, and Hector's presentation was reserved as the final presentation and thus heightened the importance and status of it. While the case could be made that the rockets and the science project are more in line with the general culture of militarism and militarization of the school and the exaltation of weapon knowledge, I argue that firing the rockets, unaided by staff or faculty, is a violent act. The students not only made actually functioning weapons (different from Shawn's masking-tape guns as his did not actually fire), but they fired them on school grounds using an air compressor. This was a violent and dangerous activity as an explosion could injure students and staff.

Beyond the science project, Peter's behavior and interactions at the school is another example of the condonement of acts of violence at the MEI. As mentioned earlier, Peter is one of the most physically mature boys at the school, as well as being one of the most aggressive. As I served as the school's noon lunchtime aide, I often watched Peter play football with 10 to 15 other boys during lunch break. Although the game was "touch" football, games regularly escalated into tackle football with Peter dominating the game by shoving, kicking, and punching other players. Sarah, the school aide, sometimes broke up the game when she decided the game had gotten "too rough," but more often than not she just yelled out threats to the group of boys and complained under her breathe about Peter's aggressive behavior. Peter's mother, Danny, came to the school everyday and stayed throughout to supervise her son's behavior. But even with his mother at the school, Peter was involved in eight different physical altercations before the end of the school year. Two of these incidents involved physically hitting two different female cadets, on two different occasions, leaving a very deep red bruise on the shoulder of one. Peter was expelled after this final incident. His expulsion occurred only after a group of mothers complained about his behavior, and the father of the victim filed a formal complaint to the school board and then removed his daughter from the MEI. Although it could be argued that this is an extreme example, or an example of bad school management, the fact is that the MEI is located within a school district with zero-tolerance policies in place. Thus, a student with eight different physical altercations (who was expelled only after pressure from parents) is indicative of the culture and acceptance of violence at the MEI. Importantly, it is also indicative of the social power and leeway granted to those who wield hegemonic masculinity in any particular context.

Violent cadet behavior, however, is not uniformly sanctioned or accepted at the MEI but runs along lines of race and gender. As Connell continually reminds us, the hierarchical power relationships of masculinity are racialized (Connell 1995, 2000), and at the MEI, violent masculine behavior is only acceptable for White male cadets. Sam, a chubby, sweet, Black seventh grader who constantly checks in on how friends, parents, and staff are doing, and generously shares his lunch and snacks with friends and is similar in stature to Peter, was suspended for several days. He was formally charged with assault after his first and only physical confrontation with another male student. Interestingly, Sam's suspension occurred within the time frame of Peter's fifth or sixth fight, thus illustrating the variable reaction by the school administration to masculine violence. Black masculinity is often linked with criminality (Welch 2007) and stereotypically seen as hypermasculine (Ross 1998; Sewell 1997) as well as overtly violent (Giroux 1996; Hoch 1979). Additionally, Black males in public schools receive harsher punishments for similar violations than their White counterparts (Ferguson 2000; Majors 2001; Schott Foundation 2004). Thus, although difference in punishment between Peter and Sam is obviously an example of racism, the similarities between the cases illustrate the way in which hegemonic masculinity at the MEI, and the use of violence, is solely reserved for White males. The examples also illustrate the fact that when "other" males try to encroach upon hegemonic masculinity, they are punished.

The controlling of masculine behavior occurs not only for males marked as "other" and males who exemplify marginalized masculinity (Connell 1995) but also across gender lines as well. Bridgette, a rowdy, White, female seventh grader, is another example of the exclusion of masculine practice in regard to violence. Bridgette and her best friend, JoAnn, spend most of their free time at the MEI running, wrestling, and play-fighting. Both girls have short cropped hair and did not display any of the illicit markers of femininity at the MEI such as makeup, jewelry, or unbuttoned blouses. The girls were social outcasts at the MEI. More often than not, when girls or women violate gender norms, they are often labeled as deviant and stigmatized (Anderson 1988; Schur 1984), and many students whispered to me that Bridgette and JoAnn were lesbians. Similarly to boys, girls also view masculinity as a source of social power (Connell 1987) and often try to capitalize upon it. Thus, when Bridgette and JoAnn sought out the advantages of hegemonic masculinity at the MEI (strength, aggression, masculine and hairstyles), their sexuality was questioned.

In the fall of the same school year as Peter and Sam's confrontations, Bridgette brought an unloaded airgun used in paintball activities to the school campus to sell it to another student, but after school hours. When the school administration learned of what was taking place, the police were called; Bridgette was arrested and immediately expelled from the MEI. Although I do not condone Bridgette's behavior, it is interesting to note that airguns use compressed air to launch projectiles such as paintballs or small pellets. This is exactly the same type of technology used in the launching of homemade rockets by the three boys in science class. Thus, while boys at the MEI can build and fire projectiles during school hours and for class credit, a girl is not allowed to bring an unloaded airgun to campus. Although it could be argued that Bridgette's case was unique and carried harsher sanctions because it was unsupervised, I argue that the differences in reaction to the actions of Bridgette and the group of boys as well as the similarities in devices and technology illustrate how masculinity and violence are condoned and accepted for White males but not for males marked as other or for females. The fact that the police were called onto school campus is an example of extreme gender sanctioning for Bridgette's role as a female trying to accomplish hegemonic masculinity at the MEI, not to mention the fact that the potential buyer of the gun, a boy, was not punished.

To sum up, violence is an important aspect of daily school life whether is it implicit in the generalized knowledge of weapons or explicit in acts of physical violence. However, only acts of violence are integral to the construction of hegemonic masculinity at the MEI. Both boys and girls tried to capitalize upon the advantages of hegemonic masculinity at the MEI, but it was only White boys who were the recipients of the advantages of condoned violence. Girls of all races and boys of color were swiftly and severely reprimanded when they tried to "cash in" on the advantages of hegemonic masculinity and thus explicit violence at the MEI. White boys, for the most part, were condoned at the MEI for the practice of violence, whether that be through schoolyard games, classroom projects, or actual physical confrontations. These condoned acts of violence are set within a generalized culture of violence at the school through the inculcation of weapon knowledge that was open to all cadets.

Girls, Masculinity, and Femininity

Masculinity is often linked to male bodies and what men do. Girls, however, can enact masculine identities and practices (Halberstam 1998; Paechter 2006) and often view masculinity as a source of social power (Connell 1987). For girls at the MEI, however, embodying masculinity is a precarious, balancing act. The MEI is a militarized and highly masculine environment. As mentioned earlier, there are aspects of masculinity at the MEI that are open to girls, but there are also aspects that are specifically delineated for boys. This section will outline how girls at the MEI negotiate the treacherous terrain of masculinity and femininity.

Tomboyism is a female masculinity that is associated with childhood and is generally accepted and condoned up until the teenage years. There are quite a few girls at the MEI who self-identify or who other students identify as tomboys. Students at the MEI describe tomboys as girls who are into "the military and drill" and who are "tough" and "strong." These "tough girls," as they were called, race and roughhouse around the school grounds during the lunch hour; more feminine, female cadets pointed them out to me with a tone of awe and respect and told me which boys they had "beat up." Tomboy narratives such as these demean traditional femininity while praising masculinity (Brown 2003). According to Halberstam (1998), such narratives are "associated with a 'natural' desire for the greater freedoms and mobilities enjoyed by boys. Very often it is read as a sign of independence and selfmotivation, and tomboyism may even be encouraged to the extent that it remains comfortably linked to a stable sense of a girl identity" (6). Halberstam's observations regarding tomboyism hold true for girls at the MEI.

For example, Maricela is one of the toughest girl at the MEI and spends the majority of her lunch break chasing down and playfully punching and wrestling the boys who are teasing her or her friends. Although Maricela is very much a tomboy, she negotiates the space between femininity and masculinity by wearing her hair long, using small amounts of makeup, and wishing the uniform for girls were "skirts or something like that." Thus, Maricela, while having a tough reputation due to her roughhousing and physical interactions with boys, still maintains her femininity through her appearance. Maricela is popular and respected by both boys and girls. Her story, however, is much different from Bridgette and JoAnn's.

As mentioned earlier, Bridgette and JoAnn spend much of their free time roughhousing and running around the school grounds, laughing wildly as they chase and wrestle each other to the ground. Both Bridgette and JoAnn have short cropped hair, neither of them wear any makeup or jewelry, and their uniforms are often dirty from their exuberant and aggressive wrestling. Bridgette and JoAnn were social outcasts. They had few friends other than each other and were speculated to be lesbians. The difference between Bridgette, JoAnn, and Maricela is that Maricela maintained a feminine identity, whereas Bridgette and JoAnn did not. Although it could seem that it is Bridgette and JoAnn's alleged sexuality that marginalized them at the school, this is not the case. There are several other girls at the MEI who were openly bisexual or queer identified but who were not ostracized. These other queer girls maintained a feminine identity through appearance, did not roughhouse, were not known to be "tough," and spent their free time listening to music and singing together. Girls at the MEI, then, may embody masculinity as long as it is circumvented by femininity, existing within the realms of both masculinity and femininity.

Thus, tomboys balance between masculinity and femininity, and girls who exhibit only masculinity are sanctioned. However, traditionally feminine girls were neither of these things. "Girlie-girls," as the cadets at the MEI called them, actively resisted the militarized and masculine culture at the MEI, embracing a traditional feminine identity instead. For example, Monica, a traditionally feminine, eighth grade girl who attends MEI at the behest of her mother, talked to me about "military stuff": "It's cool to go and watch, but it's not for me because like I'm into dancing and I'm a girlie-girl. I don't like to crawl in the mud and wear camouflage. Girls can do that, but like the girls that mainly do that are more tomboys and they are not really girlie-girls and dainty like me." Girlie-girls wore their hair long and meticulously styled, wore lots of makeup and jewelry, and unbuttoned their blouses precariously low. These girls often received uniform violations and sat in small groups at lunchtime, huddled together, talking, and watching other students' antics. Students who were heavily involved in the military aspects of the school often rolled their eyes when referring to the girlie-girls and described them as "not serious" cadets. Girlie-girls were also often chided by instructors for not "trying hard" enough during military drill practice where they would shuffle their feet and look bored, or during physical training when they would slowly jog around the soccer field giggling and chatting with each other.

Girlie-girls were not interested in participating in the military culture at the MEI and made active decisions not to participate either by consciously making mistakes during military drill, the slowing down of physical training exercises, or through exhibition of improper uniforms. These girls, for the most part, were not interested in accessing the benefits of masculinity but were active in their resistance from the official expectations and the militarized and masculine norms of the MEI.

Conclusion

Despite the large amount of work examining gender and education, as well as gender and militarization, the recent trend towards militarization of public education has gone largely unexplored in regard to gender. Schools are increasingly being locked down and supervised by police (Lipman 2003; Kupchik and Torin 2006), while students are increasingly subjected to metal detectors and the assault of military recruiters, but the effects of these trends on gender and masculinity, specifically, have largely gone without scrutiny. This chapter fills this gap by examining how masculinity is constructed at a miltarized charter school by utilizing Connell's model of hegemonic masculinity as a theoretical framework while emphasizing that masculinity is not necessarily linked to what men or male bodies do. Thus, this chapter examines militarized masculinity for both boys and girls.

Hegemonic militarized masculinity is exemplified at the MEI through the condonement of violence and the warrior hero archetype. Because hegemonic masculinity is maintained through the exclusion and subordination of other masculinities (subordinated and marginal) as well as "feminine masculinities" (tomboyism), not all cadets had access to, desired, or could capitalize upon the advantages of hegemonic masculinity at the MEI. The severe punishment and gender sanctioning for girls or Black boys who attempted to enact violence illustrates just how exclusive and illusive hegemonic masculinity is for most cadets at the MEI. The knowledge of weapons and the warrior hero archetype, however, were both much more open to girls and boys of color. While cadets never mentioned the race of the warrior hero, gender was often spoken about. Both boys and girls viewed the warrior hero archetype as something that both men and women/boys and girls could embody, although in slightly different forms. Men were argued to be stronger while women were argued to be more agile and to excel in espionage. Although women were referred to as the physically weaker sex, women were not seen to be in need of protection contrary to traditional gendered constructions. Including women and girls into the warrior hero archetype may reflect a change in gender construction for those of younger generations and the changing roles of women in the US military. However, this is not to argue that all cadets at the MEI were interested in or wanted to embody the warrior hero as exemplified by the girlie-girls and their active resistance of masculine and militarized aspects of MEI culture.

6

Ask, Tell, Talk Back: Queering Resistance to Gendered Heteronormativity

> Well, that is what me and my posse do. We talk shit about people.
>
> —Ben, a seventh grader

Sexuality is a form of power (Foucault 1990) and an organizing principle of social life. Although sexuality is usually thought of in terms of individual practices and identities, sexualized meanings and sexual boundaries are produced and enforced at institutional levels (Epstein 1994; Gamson and Moon 2004) and vary across, race, class, gender, ability, location, as well as institutional type. One of the most important institutions in the production of sexualities, especially in regard to youth, is education. School rituals, curriculum, policies, and organizational structure affirm and reinforce gendered and sexual norms. For example, as illustrated in the previous chapter, hegemonic masculinity is constructed at the MEI through the acceptance of violence and warrior hero archetype. Both Black boys and girls who attempt to capitalize on the advantages of hegemonic masculinity are sanctioned. Thus, the enactment of militarized hegemonic masculinity (particularly through acts of violence) is really only available to White boys. However, just as schools are sites of production, they are also sites of resistance illustrated by the girlie-girls who actively resist hegemonic masculinity at the MEI through uniform violations, slowing of physical training exercises, and by making conscious mistakes during drill practice. I continue along this vein in this chapter. By extending the explanatory power of resistance theory, I argue that the MEI itself is a queerly "contested space" in which gendered and heteronormative practices and identities are not only confirmed but also actively resisted by cadets at the MEI.

As discussed in detail in chapter 1, reproduction theorists conceptualize students as passive dupes who internalize and reproduce not only the material relations of the dominant group but also the social and cultural norms (Trudell 1993). Reproduction theorists tend to dismiss the agency of students and how they navigate school processes and everyday interactions of domination as well as resistance (Giroux 1983a, 1983b; Trudell 1993). Resistance theorists challenge reproduction theory by advancing that students are not only critical participants who interact with structures of domination, they are also active participants in opposing dominant forces and creating new spaces of oppositional culture and knowledge. This is particularly important in resisting and constructing gendered and sexualized meanings and identities within and against normative institutional constructions (Lasser and Tharinger 2003; Stein and Plummer 1996). Trudell (1993) refers to this as the "informal sexuality curriculum." Critical ethnographic studies have also illustrated how resistance to school structures and process is an important part of school life (Everhart 1983; McRobbie 1978; Willis 1977).

Chapter 1 discussed in depth the strengths and shortcomings of resistance theoretical models. To briefly summarize, resistance theorists have succeeded in highlighting agency and oppositional cultures within schools, but they have been much less successful at teasing out the difference between oppositional culture and oppositional culture that is motivated and shaped by active resistance to dominant school structures. Additionally, resistance theorists have failed to admit that oppositional culture may be a reaction to historical forms of domination (racism, sexism, classism, homophobia) that originate and exist outside of schools (Giroux 2001). Thus, it is important to examine the expansive social context in which resistance by subordinated groups takes place. Queer theory is a useful theoretical tool in which to further extend the important contributions of resistance theory by highlighting how school cultures are shaped by the hegemony of heteronormative and gendered practices, but also the agency and creativity that students possess to contest these behaviors and create alternative inclusive practices and space. I extend the foundations of resistance theory by integrating aspects of queer theory (heterotopic, carnivalesque moments, and subaltern counterpublics) to examine how students challenge the normative gendered and sexualized behaviors (such as sexual scripts for flirting and dating as well as homophobia and homophobic slurs) to create their own sexual rituals, practices, identities, and behaviors. In this sense, I am *queering* resistance theory.

Drawing upon queer theory, this chapter utilizes Foucault concept of "heterotopic" space (1986), Fraser's (1993) notion of "subaltern counterpublics," and Bakhtin's (1984 [1936]) concept of "carnivalesque" to examine how gendered and sexualized practices and rituals are constructed and resisted at a militarized charter school. The chapter begins with a brief discussion of schools as gendered and sexualized institutions and a brief overview of queer theory before examining how gender and sexualized norms are constructed and then resisted at the MEI. Ultimately, in this chapter I argue that the militarization of the MEI enforces normative gendered and sexualized processes and behaviors such as sexual scripts for flirting and dating as well as homophobia and homophobic slurs. While the MEI is a militarized institution of heteronormative and gendered practices, it is also a queerly resisted and contested space (through heterotopias and subaltern counterpublics) in which students actively challenge these normative standards and create their own sexual rituals, practices, identities, and spaces. The MEI is a queerly resisted and contested space.

Schools as Gendered and Sexualized Institutions

Teens, undoubtedly, spend a very large portion of their time in schools, which are highly gendered and sexualized institutions (Epstein and Johnson, 1994, 1998; Haywood, 1996; Haywood and Mac an Ghaill, 1996; Kessler et al. 1985; Thorne 1993). Institutions are not generally viewed as sexualized, but educational institutions are potently (hetero) sexualized and, consequently, constantly regulate and produce sexual meanings (Epstein 1994). Studies of sexualities in schools tend to focus on the construction of gendered heteronormativity (Robinson 2005) or marginal sexual identities and practices (Grossman 1997; Thurlow 2002; Walden-Haugrud and Magruder 1996) rather than how youth conform to and importantly resist gendered and sexualized institutional structures.

Schools are organized in heterosexual and homophobic ways (Walford 2000; Walters and Hayes 1998; Wood 1984), which cannot be separated from gendered organizational structures (Neilsen et al. 2000). Butler (1995) refers to this gendered and sexualized structure as the "heterosexual matrix" or the ways in which masculinity and femininity are ordered through a presupposed heterosexuality. Within the context of this heterosexual matrix, Lasser and Tharinger (2003) argue queer youth must decide how, when, and to whom to disclose their queer identity. Thus, identity disclosure and visibility is a process between the individual and the social-institutional context. However, Lasser and Tharinger failed to discuss how race, class, or other social characteristics affect cadets' ability to disclose their sexual identity. Additionally, Miceli (2010) argues that it is not queer youths' negative self-image but rather the negative interactions with peer and school structures that invoke negative feelings about their sexual identities. Kehily (2000) argues that schools are sites for the enactments of "heterosexual masculinities... [that] demonstrate the power of heterosexuality and the fragility of sex/gender categories" (27). Finally, Pascoe (2007) examines how masculinity is constructed through heterosexualized and homophobic social interactions and processes at a public high school in California.

Those studies that do explore sexualities in schools tend to focus on secondary or high schools even though primary and middle schools are also sites for the production of sexuality (Renold 2000; Robinson 2005). Additionally, too few studies examine how teens and youth resist and subvert institutional gendered and heteronormative processes, and not enough studies recognize that interaction in schools can be libratory experiences of power and resistance. There has also been minimal attention to the intersections of race and class, and none to my knowledge examined sexualities in a militarized public school or program. This chapter fills the gap by examining the construction and resistance of sexualities in a militarized middle and high school. I extend resistance theory by integrating concepts from queer theory (heterotopic space, carnivelsque, and subalteran counterpublics) to highlight the queer spaces and cultural practices of resistance at the MEI.

Queer Theory, Power, and Resistance

Queer theory is a useful theoretical framework in examining sexual power, knowledges, practices within schools, and particularly transgressive knowledges and practices. It is an interdisciplinary theoretical examination of sexuality and is particularly powerful in understanding how students resist and create alternative practices and identities within gendered and heteronormative institutions such as the MEI. Queer theory, however, does not create new knowledges surrounding sexualities but rather examines how sexual knowledges and categories are created and contested. Stein and Plummer (1996) laid out the basic principles of queer theory: "(1) a conceptualization of sexuality which sees sexual power embodied in different levels of social life, expressed discursively and enforced through boundaries and binary divides; (2) the problematization of sexual and gender categories, and of identities in general; (3) a rejection of civil-rights strategies in favor of a politics of carnival, transgression, and parody which leads to deconstruction, decentering, revisionist regains, and anti-assimilationist politics; and (4) a willingness to interrogate areas which normally would not be seen as the terrain of sexuality, and to conduct queer 'readings' of ostensibly heterosexual or non-sexualized texts" (134).

Thus, from a queer theory perspective, sexual identities are contingent, created, and recreated discursively through social interaction of parody and transgression. Sexual identities and categories are fluid and shifting (Esterberg 1996). Sexual identity is not something one discovers about one's self, but produced and reenacted through behaviors and interactions (Plummer 1996; Seidman 1996). Queer theory additionally illuminates sexual power in areas of social life that are usually not viewed as sexual such as cultural forms and social institutions like the economy and education (Seidman 1996). The specific organization of any institution can shape and (re)produce sexual knowledge and practices. What is more, queer theory argues that gender and sexual identities are mutable and mediated through power relations. Power exists in all relationships (Foucault 1990) and comes not only from above but is also exercised from below. Resistance is inherent in all power relationships, not external to them, and in fact bolsters the dominant discourse by clarifying boundaries and legitimacy. By centering power, queer theory provides a conceptual lens through which to analyze resistance to hegemonic gender and sexual norms.

Resistance, Carnivalesque, and Subalteran Counterpublics

Resistance is a source of power for marginalized youth (Blackburn 2004; Foucault 1990) and even small acts of resistance against heteronormativity can be powerful for queer youth (Letts 2006). This is particularly true within schools, which are laden with heteronormative practices and norms. Several studies have highlighted how students resist and transgress heteronormative gendered school structures and cultures. For example, Dalley and Campbell (2003) found that a group of girls known as "The Nerds" at a high school in Canada were able to resist heteronormative gender standards (female students as passive and sexually attractive) by taking on public lesbian personas, even though the girls did not actually consider themselves to be lesbian. Additionally, Rasmussen (2004) utilizes the Foucauldian notion of "heterotopia" to examine how particular spaces can be locals of heterosexuality while simultaneously being arenas in which designated sexual functions are subverted or neutralized. Heterotopias are real spaces that serve as countersites that contest and invert the original designated function of that space (Foucault and Miskowiec 1986). Foucault argues that in order to escape everyday oppression and control, heterotopic spaces must be created in which difference can be celebrated and enacted. For example, school dances are traditional heterosexual spaces that can be neutralized when students attend in large mixed-sex groups rather than as heterosexual couples. Finally, Blackburn (2004) examines how queer youth use language to subvert oppression and homophobia into experiences of pleasure. Blackburn argues for educators and researchers to acknowledge that there are different and simultaneous discourses students utilize to position themselves as agents rather than victims.

Bakhtin's (1984 [1936]) concept of "carnivalesque" and Fraser's (1990) "subalteran counterpublics" offer two additional powerful analytical tools with which to queerly examine power and resistance within heteronormative and gendered spaces. Carnivalesque is a subversion of the dominant hierarchical social order through humor, parody, and chaos (see the third characteristic of queer theory above). For example, Atkinson and DePalma (2008) argue that the success of the girls in Dalley and Campbell's (2003) study (discussed above) to resist heteronormative gender standards was due to the fact that these girls were "burlesquing" gender and sexuality norms. Atkinson and DePalma (2008) define burlesquing as "a larger than life purposeful inversion of social, gender-related mores"(30).[1] They argue that it is within playful carnivalistic moments such as these that "momentary reversal of conventions" (31) arise. Thus, burlesquing, through acts of parody and play, can subvert hegemonic practices such as heteronormativity.

The second useful analytical concept in understanding resistance within heteronormative and gendered institutions is Fraser's "subaltern counterpublics" or "parallel discursive arenas where members of subordinated social groups invent and circulate counterdiscourses, which in turn permit them to formulate oppositional interpretations of their identities, interests, and needs" (67). That is the creation of alternative spaces by marginalized groups, out of exclusionary necessity, to counter stereotypes and assert self-defined identities and interests. For example,

Weis and Fine (2001) applied the concept in public schools illustrating how educators created parallel yet oppositional public spaces to contest hegemonic constructions of gender and race. For example, by decentering privilege in a public school program, the educators in Weis and Fine's (2001) study facilitated a space in which young women critically challenged gender and racial stereotypes. Additionally, Wells (2006) examined how three, male gay student activists challenged exclusionary public educational spaces through "subalteran counterpublics" (e.g., starting a positive/safe space campaign in high school) and became agents of social change within their schools. In this chapter, I draw on queer theory, specifically Foucault's physical concept of heterotopic space, Fraser's discursive concept of "subaltern counterpublics," and Bakhtin's concept of "carnivalesque," to examine how heteronormative gendered and sexualized practices and rituals are constructed and resisted at a militarized charter school by creating alternative public space through humor and parody. I argue that the MEI is a "contested space" of normative gendered and sexual practices and identities as well as a creative space in which students resist (physically and discursively) such practices at the MEI.

Heteronormative, Gendered Practices, and Resistance at the MEI

Schools are institutions that reinforce heteronormative gendered norms and standards. While the MEI is no different in this aspect, it is unique in that certain aspects of the heteronormative, gendered standards reflect the militarized structure of the school. Heteronormative and gendered practices are created and reinforced in two main ways at the MEI: first, through heteronormative sexual scripts of flirting and dating, and second, through homophobia and homophobic slurs. Cadets, however, are not passive in the construction of sexual and gendered norms. They are active participants in both production and transgression of such norms. The following sections will detail the construction of heteronormative and gendered practices at the MEI and the ways in which cadets resist and transgress these norms through carnivalesque moments and subaltern counterpublics. I will first outline the heteronormative sexual scripts for flirting and dating (public affections, flirting, and the military ball) and how students contest these followed by a section detailing homophobia and homophobic slurs and how such practices are challenged.

Heteronormative Sexual Scripts of Flirting and Dating

While the MEI is a highly structured and fully militarized educational institution, it should not be forgotten that the cadets who attend the MEI are teens navigating the often contested and complicated terrain of sexuality. On several occasions, cadets asked me questions about sexuality, usually questions regarding STIs. The MEI did offer a day of sexual education, but approximately 15 cadets did not have parental permission to attend. I was asked to sit with these cadets during the class when sex was being discussed, and these cadets were genuinely curious about what was being taught. When I explained to them that there was an entire chapter in their biology textbooks on human sexuality, the cadets hurriedly pulled out their textbooks and flipped to the chapter and began to devour the pages, stopping to ask me questions or clarifications. Some of the questions arose from information in the book and some arose from the student's general lack of knowledge about sexual matters.[2] It is not just during sex education classes or when reading the back chapters of biology textbooks that MEI cadets are interested in sex and sexuality. The hallways and lunch periods at the MEI are filled with the chatter of who likes who, who is dating who, and the daily analysis of online teen flirting and dating drama. With or without parental permission, the cadets at the MEI are interested in sex and interested in each other.

Gagnon and Simon (2005 [1973]) argue that sexuality is not based solely within the confines of biology, but is a social, historical, and cultural phenomenon. There are appropriate sexual behaviors and practices that are bound in cultural location and learned through social interactions that follow predictable "sexual scripts."[3] Many of these sexual scripts delineate behavior for heterosexual adolescent flirting and dating and are bound within traditional gender norms (Gagnon 1990; Schwartz and Rutter 1998) such as men should initiate sex, and the sexual double standard that provides men with more sexual freedom and agency than women (Blanc 2001). Heteronormative and gendered cultural-level scripts are evident in three main ways at the MEI: public affection, flirting, and end-of-the-year military ball (the MEI equivalent to a senior prom).

Public Displays of Affection

First, while not officially allowed by school rules, publc heterosexual interactions are common at the MEI. The public heterosexual displays of affection by students simultaneously reinforce heteronormative sexuality and defy it as students violate norms of sexual experience (i.e., premarital sex). Heterosexually coupled cadets spend time together during the lunch hour holding hands and talking quietly while sitting atop lunch tables; kisses and quick hugs are stolen between classes when teachers, parents, and cadet military police (MEI's version of hall monitors) are absent or distracted. Heavy make-out sessions and other sexual activities are common at the MEI, during school hours, after school, and on weekends. Yssenia, a friendly and confident eighth grader, took it upon herself to constantly update me on the sexual activity at the MEI and often filled me in during the lunch break of the sexual activity of her peers. Much of this sexual activity was out of sight of parents, teachers, and administrative staff, within peer-defined safe spaces such as adult-free homes, public parks, and—for those cadets who could drive and had access to—cars. Although the majority of sexual activity at the MEI was away from adult supervision, it was witnessed by and publicly known by cadets at the MEI. As Christina, a Latina ninth grader, put it, "When I came from Catholic school to [MEI] I was amazed by how many kids were already having sex. They were like, 'you haven't done it yet?' and I was like 'no'!"

Not all sexual activity went without adult acknowledgement. During my time at MEI, there were several instances when heterosexual couples were engaged in varying degrees of sexual contact from making-out to breast and genital fondling. One morning as I quickly came around a hallway corner, I happened upon a heterosexual couple that was involved in a deep and absorbing make-out session. They disentangled themselves and hurried way. When I mentioned this to the administrative assistant, she rolled her eyes saying, "They got caught the other day too. Somebody is going to get pregnant."

Parents were also aware that their teens were sexually curious and experimenting. The morning before the military ball, several parents came up to me at different times requesting me to make sure the cadets refrained from "dirty dancing" as I was one of the chaperons for the dance. As one parent I had never met and who introduced herself as a mom said, "Brooke, there can't be any dirty dancing. I know dancing isn't like how we used to dance, but they can't do *any* dirty dancing."

Public heterosexual displays of affection (whether within adult acknowledgement or not) were common at the MEI although not necessarily completely condoned. Public heterosexual affection was accepted within limits by the parents, teachers, and staff at the school, but the undercurrent of sexual energy and sexual curiosity was running strong around and behind the watchful gaze of adults. Thus, heterosexual students who were sexually active were conforming to heterosexual norms of dating, but were also defying other sexual norms that limit the amount of heterosexual intimacy and sexual experience deemed acceptable between students.

There were also same-sex displays of affection and sexual interaction, and much like the more intense public heterosexual displays of affection, much of the same-sex affection was beyond the surveillance of adults. To be honest, I never witnessed same-sex public displays of affection. For example, Yessenia (mentioned earlier) explained to me how "annoying" the new seventh grade cadets were because "they cut themselves and then they are all bi and ugh... and they are making out." When I asked who were making out, she named several seventh grade girls. While this all could just be hearsay and rumor that is common at all schools, it is important to note that these rumored girl-on-girl make-out sessions were frequent and implicated different girls at different times. That is, the rumors were not about the same girls over and over, but included different girls across grade levels. Adults, myself included, were not privileged to these moments of same-sex sexual interaction as these sexual moments were solely within the spaces for and created by youth.

The girls—and it was only girls—who were implicated in such rumors were students who were marginalized at the school. They were not particularly popular; they did not succeed in any of the extracurricular activities at the school such as military drill or sports, nor were these girls top academic performers. They were not worst at any of these activities either, but existed in the margins of the MEI. All of the girls implicated in same-sex rumors were girls of color (both Black and Latino) and feminine performing, except for Bridgette and JoAnn who were both White and performed a more masculine identity with short hair, roughhousing, and lack of makeup or jewelry. What these moments of same-sex sexual interaction created was a figurative space, a parallel space, in which to resist and challenge the heteronormative and masculine structure of the MEI. These girls, I argue, resist these structures through with their physical bodies (brown and black) and their queered sexual actions. The girls created a physical space of heterotopia not only to resist the culture of the school but also to create different and inclusive spaces in which to enact their own sexual identities and practices. Additionally, the queer rumors that circulate about these girls create a discursive, subaltern counterpublic that challenges the heteronormative sexual identities at the MEI.

Flirting

Sexual scripts for flirting are based on heteronormative and traditionally gendered scripts. Like many schools in the United States, such heteronormative and gendered sexual scripts for flirting and dating exist at the MEI. However, there are also alternative flirting scripts that students utilized to actively challenge heteronormative and gendered norms at the MEI. Cadets created a subaltern counterpublic to normative flirting culture in which to create and enact same-sex desire and identities.

To start, boys at the school overwhelmingly ask girls out on dates to attend various dances and military balls and initiate sexual/dating interactions. Additionally, both boys and girls "asked out" other cadets of the same sex. For the majority of cadets, asking another cadet "to go out" was akin to being a couple and would often lead to meeting up during the lunch break and hanging out after school. Such couplings could also lead to public displays of affection, but this depended both on the couple and the age of the couple as older cadets tended to participate more heavily in public affection. Both boys and girls asked cadets of the same sex to "go out." For example, Bridgette eventually asked out Christina (from above), and Ben asked out one of the most popular athletic boys at the MEI, Bryan. When Ben told me he had asked Bryan out, I was a bit concerned about Ben's safety at the school (see the section below on homophobic slurs and teasing). I asked Ben a few questions regarding Bryan to get a sense of how the situation played out.

> *Ben*: I asked out Bryan today.
> *B.J.*: Sporty Bryan?
> *Ben*: mmm...hmmm
> *B.J.*: Is Bryan interested in guys?
> *Ben*: mmm...hmmm
> *B.J.*: How do you know?
> *Ben*: I have my sources.
> *B.J.*: Like what?
> *Ben*: I can't tell you...I just know.
> *B.J.*: So what did he say?
> *Ben*: He said no.

Although Ben was not successful in "going out" with Bryan, he was not ridiculed or bullied for asking out one of the most athletic and popular boys at the school. Bryan communicated a simple no to Ben, and Ben turned his attention to Eddie who he had "the hots for." Thus, there was a discursive space, a subaltern counterpublic, for same-sex flirting and dating at the MEI, and while this was not successful for Ben, it did not result in harm or ridicule either.

Additionally, cadets actively flirted within the same sex and transgressed traditionally gendered scripts for dating and flirting through "stalking." Stalking is when a cadet learns the schedule of another cadet who they like and coincide their activities and times in hopes of bumping into that cadet. It is common knowledge when someone is being stalked or stalking someone. It is a passive aggressive form of flirting and is particular to girls flirting with other girls at the MEI. Two of the most popular and highest in military command at the MEI were "stalked" by other girls and asked out. Christina (mentioned earlier) explained to me one afternoon that she and her friend Amanda were currently being "stalked." "My friend Amanda has a girl who was stalking her too. Bridgette and then Marina is after me... She like stalks me in the bathroom. Yes, she knows my whole schedule! She would go to the bathroom every five seconds just to see if I was going to the bathroom."

Much like public displays of affection, same-sex stalking occurred in physical spaces where adults usually were not, such as school hallways and bathrooms. Boys, to my knowledge, did not participate in stalking nor did stalking occur between boys and girls. Stalking is a way for girls at the MEI to express same-sex desire outside and beyond the gaze of adults. These girls used the physical space of the school (hallways and bathrooms) in which to create and enact practices of same-sex desire and create a heterotopic space. I think it is important to go into some detail of the physical structure of the MEI as, I argue, it aided the construction of contested space for the practice of girl stalking.

At its second location, the MEI was a jumble of modular classrooms that were connected through hallways. These hallways were not part of the original modular buildings, but constructed later by the school district to physically connect different classrooms. The physical ground or campus at this location was not fenced off or enclosed in anyway. Additionally, the modular buildings sat adjacent to a city park. Thus, the public could easily enter the school grounds and interact with MEI cadets. The school district enclosed the modular classroom buildings as a measure of Foucauldian discipline (1975) in which to control the physical movements of students and possible (dangerous) interactions with the public. What resulted, however, was a maze of dark and narrow hallways with constant rights and lefts. The longest hallway was possibly ten-foot long. The physical construction of the hallways encouraged the practice of stalking as there were many corners and dark spaces in which to watch, but not necessarily be watched. Girls who participated in stalking utilized the physical construction of the school to stalk other girls, resist heteronormative practices of flirting, and create a physically bound heterotopic space in which to enact same-sex desire.

Chasing games are also common forms of flirting scripts, especially among the younger seventh and eighth grade cadets. At the MEI, chasing games consisted of boys initiating some teasing or annoyance to a group of girls followed by one or two of the girls from the group retaliating by chasing the boys and "beating them up," as Maricela, a fierce seventh-grade Latina who planned on joining the military, explained. This usually resulted in the boy being wrestled to the ground and punched playfully in the arm and chest. The girls who participated in such heteronormative chasing games were interested and active in the military aspects of the school (part of the competitive military drill team and morning flag ceremony). These girls were known at the school as tomboys and self-appointed "tough girls." Interestingly, even though this group of girls identified themselves as tomboys and tough girls, they maintained a feminine gender performance keeping their hair long and curled and wearing trace amounts of makeup (such as clear lip gloss) and jewelry (stud earrings or a thin necklace). Thus, these tomboys and "tough girls" are keeping in line with the gendered heteronormative scripts of teen flirting and female appearance, but are also defying gendered norms by assertively wrestling and chasing boys.

Finally, there are cases of girl-on-girl, sexually charged chasing games. As mentioned earlier, often cadets participated in heterosexual flirting involving chasing games in which boys would annoy a girl or group of girls and the girls would react by chasing away the boy or group of boys. Chasing games were also common among girls at the MEI, and Bridgette and JoAnn were the most rambunctious participants in same-sex chasing games. As mentioned in chapter 5, Bridgette and JoAnn spent much of their free time roughhousing and running around the school grounds, laughing wildly as they chased and wrestled each other to the ground. Neither of the girls was gender-normative as they both had short cropped hair, never wore any makeup or jewelry, and their uniforms were often dirty from their exuberant and aggressive wrestling. They were social outcasts at the MEI. They had few friends other than each other and rumors about sexual identities of the two swirled around the MEI.

The wrestling that Bridgette and JoAnn participated in was not fundamentally different from the chasing and wrestling other cadets participated in, but it was more prolonged and focused. Whereas when boys and girls participated in such flirting and chasing games, the members and quantity of the groups changed daily as different boys instigated the games with different girls. The chasing between girls and boys was also shorter, ending with one of the girls playfully punching one of the boys. Bridgette and JoAnn were different in that they only chased each other and took turns being pursued. Their chasing games lasted much longer than the boy/girl games and were played with much more intensity. JoAnn would chase Bridgette until she wrestled Bridgette to the ground upon which JoAnn would run off with Bridgette closely after her. Although it could seem that it is Bridgette and JoAnn's alleged sexuality that marginalized them at the school, this is not the case. There are several other girls at the MEI who were openly bisexual or gay and who were not ostracized as Bridgette and JoAnn were. These other queer girls maintained a feminine identity through appearance, did not roughhouse, were not known to be "tough," and spent their free time listening to music and singing. Chasing games are a common flirting script among teens, and the cadets at the MEI are no different. However, chasing games at the MEI were not only boy/girl groupings but also girl/girl groupings, allowing for same-sex flirting to exist within physical contact and play. The boy/girl chasing games simultaneously challenged and reinforced gendered and heteronormative sexual practices. While girl/girl chasing games were present at the MEI, the girls who participated in these were contesting gender and heteronormative sexual practices and creating heterotopic spaces in which to enact same-sex desire. However, due to their explicit defiance of feminine gender norms, in terms of appearance, they were socially sanctioned.

Military Ball

Lastly, the military ball exemplifies the heteronormative and gendered sexual scripts at the MEI, but also holds the creative possibility in which to challenge and resist them. The military ball is similar to a formal dance such as senior prom common in US high schools. However, the military ball at the MEI required all cadets to wear their "dress blues" or the MEI uniform, which consists of dark blue slacks, blue button-down shirt with the embroidered school logo, shined black shoes, and polished belt.[4] All cadets were allowed to attend the ball, with or without dates, as long as they were in the correct uniformed attire. The ball was held in a nearby city gymnasium (the school did not have its own athletic facilities at that time) and was organized and decorated by a group of parents who were active at the school. Cookies and punch lined one wall of the gymnasium and a local DJ blasted popular music and encouraged the cadets to dance and sing along. Surrounding local JROTC and other private militarized schools and programs were invited to attend. While some MEI cadets attended together with dates, most cadets attended the ball in large, friend-based groups. The DJ played mostly hip-hop and top 40 hits. The cadets danced in large groups, jumped up and down to the beat of the music, and screamed the popular lines from the songs of artists such as Rihanna and Beyoncé. While some cadets were paired up in heterosexual couplings either with dates or for slow dances (which were not popular and cleared the dance floor), the vast majority of cadets danced and interacted in large groups that contained both boys and girls.

The military ball appears at a first glance to be steeped in heteronormative scripts; however, the military was a site of active resistance of gendered and heteronormative practices. The military ball was, in the Foucaldian sense, a heterotopic space. There were three main ways in which the military ball transformed from a heteronormative space to a space of resistance: the ability for students to attend as couples, in groups, or alone; the lack of gendered formal attire; and finally the dancing patterns and behaviors of the cadets.

First, cadets were allowed to attend the ball as a couple, alone, or in groups. Cadets did not have to have a date or a heterosexual pairing to be able to attend. Only a few cadets attended the ball with a date and those who did attended in a heterosexual pairing. Most cadets attended with large friend-based groups, and many cadets were dropped off and picked up at the dance by parents or guardians. This lack of focus on heterosexual pairings allowed for the behaviors and interactions of the cadets to be less constrained by heterosexual scripts for flirting and dating as well as for interactions to be more fluid.

Second, as mentioned earlier, cadets at the MEI and neighboring military schools could attend the ball as long as they wore a school uniform. There was much contention over this rule as many older MEI cadets wanted to attend the ball in traditional formal attire such as tuxedos and ballgowns. However, the MEI administration decided that mandating that all cadets wear the militarized school uniform to the dance would alleviate any cadets not being able to attend due to financial stress as many of the cadets who attend the MEI are working-class or poor. While the goal of this rule was to be financially inclusive to the cadets at the MEI, it created a unique atmosphere in which gendered clothing was not allowed, as the uniform is the same for both male and female cadets. Thus, the military ball was a space in which traditional gendered norms in terms of dress at school dances were leveled as all cadets attended in nonspecific gendered uniforms.

Finally, dancing patterns and behaviors of the cadets were observed. As most cadets did not attend the ball with a date, cadets overwhelmingly danced in large friend-based groups rather than as heterosexual couples. This style of dance is partly attributable to the fact that the DJ played popular, fast-paced music, as on playing slow music, the dance floor cleared and cadets stood around looking bored. One interesting aspect of group dancing was the sexually charged dancing by girls with other girls. There were more girls dancing with other girls in sexually suggestive way than between boys and girls. Girls danced close to each other, grinding their hips while tossing their hair side-to-side. They bent over and shook their backsides at each other and mimicked lap dancing until the chair was sternly taken off the dance floor by Mr Jones, one of the social studies teachers at the MEI and a retired navy veteran. The dancing was sexually charged, and both straight and queer-identified girls participated in this type of dance.

Dance is a type of sexual expression for young women, and as young women have few choices for sexual expression, dance is one that allows women personal pleasure and power within sexuality (McRobbie 1984), and in this case personal pleasure and sexual power by and between girls. Other than the removal of the chair by Mr Jones, this type of dancing was never overtly critiqued at the MEI. Parents and chaperons at the ball only intervened if the dancing was sexually charged between a boy and a girl, not, interestingly, between girls. This is particularly intriguing as the dancing violates norms surrounding sexuality (as girls should dance with boys), but simultaneously reinforces gendered norms (girls dance). Girls were not scolded for dancing in this particular manner, nor were they asked to stop. It could be that girls are allowed to display more physical (nonerotic) intimacy with other girls such as touching, hugging, etc. It could also be likely that the teachers and parents were so concerned about heterosexually charged dancing that they completely missed the undercurrent of sexual energy of girls dancing with other girls.

While the explicit intentions of these girls dancing with each other are unknown, the dancing is, I argue, an example of sexual agency and bodily pleasure for female cadets at the MEI. The heteronormative alternative would be for female cadets to wait to be asked to dance by male cadets. This was not the case and only heterosexual couples who attended the ball together danced in heteronormative gendered ways. What is fascinating about all of this is the fact that a military ball is the bastion of gender and heteronormative flirting and dating scripts. It is a space designed for the preservation of gendered norms and heterosexuality. However, in this instance, it is a space of resistance, a space of sexual rebellion or what Foucault (1986) refers to as a "heterotopia" or "counter-sites…in which real sites that can be found within the culture, are simultaneously represented, contested, and inverted" (24). Such spaces are capable of positioning in one real space several spaces that are incompatible with one another. Thus, the military ball is at once an institutionalized space that reinforces gender and heteronormative sexual practices while it is simultaneously a space to resist these practices through dance, clothing, and group attendance.

Interestingly, many cadets stated that they wanted to transfer to "a real high school" to be able to play sports and to attend a more traditional end-of-year dance such as prom. Even parents were aware that the lack of sports and proms was a significant negative factor for their children. For example, Susan Uppal, a highly active parent at the MEI, explained why she would send her daughter to a "traditional" high school:

> My daughter won't go to high school here. She is going to go to a traditional high school and that is because I promised her that if she did junior high here then I would let her go to a traditional high school if she wanted to at some point. She has expressed that she wants to…She wants to be involved in all the things, the proms.

In the end, the military ball is a site of slippage, a contested space. The military ball is not a fully institutionalized heteronormative, gendered space, but a Foucauldian heterotopic space to resist and transgress such heteronormative and gendered processes. This heterotopic space allowed for nonheteronormative social groupings as cadets attended in friend-based groups and in nongender-specific uniforms as well as nonheteronormative bodily and physical expressions of pleasure and desire.

Homophobia and Homophobic Slurs

Homophobia is intricately linked with the social construction of masculinity (Kimmel 2003 [1994]), and authors have additionally pointed out the historical and reciprocal ties between the military and the construction of masculinity (Goldstein 2001; Higate and Hopton 2004). As the US military is a highly masculine institution, homophobia is also rife within it and was formally institutionalized (Belkin 2001) until the ban on gays and lesbians serving in the military was lifted in September 2011. Homophobia is also a common aspect in school cultures (Burn 2000; Miceli 2010; Pascoe 2007; Plummer 2001; Smith 1998; Walters and Hayes 1998).

The MEI is unfortunately not unique in that homophobia is a common aspect of the school culture and is exemplified in three main ways: the use of the phrase "that's so gay"; rumors regarding cadets' sexual practices and identities; and finally gay bashing and teasing. To begin with, much like other schools and educational institutions, the phrase "that's so gay" was ubiquitous at the MEI. Seemingly anything the cadets disliked at the schools was termed "gay." This included rules, uniforms, behaviors, other cadets, teachers, and jokes; there was seemingly an unending list of people, actions, and things that were "gay." When I first began my research at the MEI, I was, somewhat naively, surprised that such terminology was still in use as much as it was in use in my middle and high school days in the 1990s. I was additionally surprised by the fact that even queer cadets used the terminology especially in regard to the school. For example, Ben used "gay" when he was telling me how much he disliked attending the MEI. "I just hate it. Bottom line. I hate everybody in it except my clique. It's gay. It's stupid. I always get into trouble." This practice of devaluing sexual practices and identities that fall outside of heterosexuality is what Warner (1993) refers to as "heteronormalizing practices," or practices that construct heterosexual behaviors as natural and normal. Thus, the ubiquitous language of "that's so gay" at the MEI, by both straight and queer-identified students, is a homophobic cultural and linguistic norm that privileges heterosexuality.

Second, cadets at the MEI were in constant conversation of who is dating whom, who likes whom, who has and has not had sex, which couples broke up and who hooked up, and who is gay. As mentioned earlier, there were constant rumors about the sexual identity of Bridgette and JoAnn. Neither of the girls was openly queer-identified, but Bridgette was more vocal about her "crushes" or who she liked often claiming that a certain boy or classmate "stole her girl." Jackie and her friend, Linda, were also rumored to be bisexual, and these rumors stemmed from online activity between the two girls on a social networking site. There were also rumors that Ben and Eddie were gay. While Ben has a limited open bisexual identity (he was openly identified so with select students and adults), Eddie identified as straight. While I could go on and on about the rumors of sexual identity at the MEI, the important thing to note is that rumors such as these were not positive. Such rumors were often whispered to me during break times and shrouded in shame and homophobia.

More important than the factual basis of the rumors is that rumors about sexual behaviors and identities often lead to the second exemplification of homophobia at the MEI: gay bashing and teasing. Such gay bashing and teasing of cadets occurred, often based on rumor, whether the cadet self-identified as queer or not. While I heard about such bullying and teasing at the MEI from cadets, I rarely witnessed it. However, on one afternoon, I came across a small group of cadets teasing Ben about being gay and being in some sort of emotional or dating relationship with Eddie (another Latino male rumored to also be gay). The teasing subsided once I intervened, but it was clear that this was not an unordinary social interaction, but something that happened with regularity beyond the bounds of adult supervision. I used this moment to have a discussion of homophobia and discrimination with the cadets involved, and a week later, Ben pulled me into a broom closet to blurt "I'm bi!" with teary eyes. While this moment of resistance to homophobia was led on my part, cadets at the MEI were actively resisting homophobia at the MEI.

Dalley and Campbell (2003) found in their ethnographic study of a high school in Canada that a group of girls took on public lesbian personas as a way to challenge heteronormative gender standards even though they did not consider themselves to be lesbian. The girls used this lesbian identity or "counter-hegemonic sexual persona" (25) in order to challenge the passive female gendered stereotype that was prevalent at the school. Students at the MEI also take on counterhegemonic sexual personas in order to actively resist heteronormative and gendered culture of the MEI. This group consists of a group of Black girls and Ben. This group often spent their free time dancing, clapping, and singing along to popular songs. This group was one of the most vocal and boisterous groups at the school. Ben often referred to this group as "his clique," "his crew," or "his posse." This group, while singing and dancing together, often shouted loudly that they were bisexual—a very in-your-face declaration of sexuality. I came across Candace, a particularly bubbly and boisterous Black seventh grader, shouting loudly as cadets streamed into the school after lunch "Oh yeah! We're gay! Oh yeah!" This group was not particularly popular at the MEI, nor was the group highly involved in any of the military command structure of the school or extracurricular activities such as the military drill team or sports teams. This group was also not ostracized, as were Bridgette and JoAnn, but was not wildly popular either.

The group did have an oppositional stance toward the school utilizing rumor and sexual identity to create an alternative space to the White, hegemonic masculine military culture of the school. When discussing the school with Ben one afternoon, he told me "everyone is mad at me right now [for doing] just what I do everyday. Talk about people. Well, that is what me and my posse do. We talk shit about people." This "talking shit" was a way for this group to resist the structures of the school. While I never enquired about the sexual identities of the members of the group, their public identities as bisexual allowed this group to challenge the heteronormative and White military and masculinist structures of the school. Ben's posse utilized queer identity in much the same way at the MEI as the girls in Dalley and Campbell's (2003) study. This group of Black girls and Ben, a Latino, could not access heteronormative sex and gender constructed as White, militarized masculinity (discussed in chapter 5), but the group was able to resist gender and heteronormative norms through public displays of bisexual identity or "counter-hegemonic sexual persona[s]." Ben's posse used song, dance, and humor to subvert the dominant hierarchical and heteronormative social order at the MEI. Through the actions of this group, the MEI became a carnivalesque and subaltern counterpublics in which to resist the militarized and heteronormative structures of the MEI and create alternative practices, behaviors, and identities. They used song, dance, and in-your-face claims of sexual identity to challenge and contest the cultural practices at the MEI. The resistance was loud, boisterous, and fun. They talked back, made fun of, and laughed at the normative and controlling structures at the MEI and created alternative, subaltern counterpublic spaces of identity and inclusion.

Conclusion

The militarization of the MEI enforces normative gendered and sexualized processes, identities, and behaviors through rituals and sexual scripts of public displays of affection, flirting, and the end-of-the-year military ball. Cadets at the MEI were active participants in the production of such norms as well as to resist and transgress such structures and practices.

To begin with, public heterosexual displays of affection, while not sanctioned at the school, were known about by parents, teachers, and staff and tolerated. However, while public heterosexual displays of affection were conforming to heteronormative and gendered scripts, students were simultaneously violating norms surrounding the amount of sexual experience acceptable (e.g., premarital sex). There were also moments of public displays of affection by same-sex couples that were beyond and outside of the watchful eyes of adults at the MEI. Interestingly, it was only feminine performing, girls of color who participated in same-sex public displays of affection. Thus, these girls at the MEI used their bodies and same-sex public displays of affection to combat the hegemonic and heteronormative construction of militarized masculinity by creating heterotopic spaces of resistance.

Second, there were heteronormative and gendered sexual scripts for flirting and dating at the MEI. Boys at the school overwhelmingly ask girls out on dates to attend various dances and military balls and initiate sexual/dating interactions. Students also "asked out" other same-sex students. While these same-sex attempts were often not successful, they also did not result in harm or ridicule. Thus, there is a space, a discursive subaltern counterpublic in which cadets expressed same-sex desire by asking other students out.

Cadets at the MEI also transgressed traditionally gendered heteronormative scripts for dating and flirting by same-sex "stalking." Girls who participated in stalking utilized the physical construction of the school (hallways and bathrooms) to stalk other girls, resist heteronormative practices of flirting, and create a physically bound heterotopias in which to enact same-sex desire compared to the discursive subaltern counterpublic space of flirting scripts.

Chasing games are also common forms of flirting scripts as boys chase particular girls and the girls retaliate by chasing the boys. These types of practices are in line with the gendered heteronormative scripts of teen flirting that occurs across schools. However, chasing games at the MEI were not only boy/girl groupings but also girl/girl groupings, allowing for same-sex flirting to exist within physical contact and play. The boy/girl chasing games simultaneously challenged and reinforced gendered and heteronormative sexual practices. Girl/girl chasing games challenged gender and heteronormative sexual practices, but due to these girls' explicit defiance of feminine gendered norms, in terms of appearance, they were socially sanctioned at the MEI. Even though those girls who participated in girl/girl chasing were socially sanctioned for opposing feminine gendered norms, they were also simultaneously creating a physical space(s), a heterotopic space, in which to enact same-sex desire.

Finally, the military ball, which appears at a first glance to be seeped in heteronormative scripts, is a site of active resistance of gendered and heteronormative practices. The military ball is, in the Foucaldian sense, a heterotopic space. Cadets at the MEI transformed the military ball from a heteronormative space to a space of resistance and creativity through the ability for students to attend as couples, in groups, or alone; the lack of gendered formal attire; and finally the dancing patterns and behaviors of the cadets. The lack of focus on heterosexual pairings allowed for the behaviors and interactions of the cadets not to be constrained by heterosexual scripts for flirting and dating as well as for interactions to be more fluid. The military ball is a space in which traditional gendered norms in terms of dress at school dances is leveled as all cadets attend in nonspecific gendered uniforms. Additionally, cadets overwhelmingly danced with each other in large friend-based groups rather than as heterosexual couples. Girls were dancing with other girls in sexually suggestive ways more often than between boys and girls. The dancing was sexually charged, and both straight and queer-identified girls participated in this type of dance. The girl/girl dancing violates norms surrounding sexuality (as girls should dance with boys), but simultaneously reinforces gendered norms (girls dance). Girls were not scolded for dancing in this particular manner, nor were they asked to stop. Thus, the military ball is a heterotopia, a physical space, in which cadets not only resisted social norms surrounding gender and sexual practices but also created and shared new ways of enacting same-sex desire and practices.

Unfortunately, the MEI is not unique in that homophobia is a common aspect of the school culture and militarized masculinity present at the school. Homophobia at the MEI is exemplified by the use of the phrase "that's so gay," rumors regarding cadets' sexual practices and identities, and gay bashing and teasing. However, a group of boisterous students actively resisted the homophobic and heternormative militarized masculine culture of the school through the enactment of queer identities through play, laughter, and "talking shit." The school serves as a carnivalesque (Bakhtin 1936) space for the subversion of the dominant hierarchical social order though humor, parody, and chaos exemplified by the loud and boisterous group of vocally bisexual cadets who created an in-your-face subaltern counterpublic.

In sum, the MEI is a militarized school where normative gendered and sexualized social practices, rituals, identities, and behaviors are enacted. However, it is also a contested space in which students actively resist these gendered and heteronormative standards through carnivalesque moments, creation of physical heterotopias and discursive subaltern counterpublics as well as material and immaterial spaces to create alternative and inclusive sexual rituals, practices, identities, and behaviors. The students at the MEI successfully, I argue, queer the physical and discursive space of the MEI.

7

Conclusion

On a crisp spring Friday morning, cadets at the MEI just finished morning formation and the "Daily Dozen," a routine of 12 different calisthenics. They are dividing into teams for an intense game of dodge ball: Third Platoon versus Second Platoon. The game begins with Mr Jones shouting out the school motto: "Stay Alert! Stay Alive!" Cadets race to the center of the concrete court scrambling for the red rubber balls. As the balls are released into the air, cadets yell "Fire!" With four balls zigzagging back and forth across the small court, it becomes a frenzied blur of red balls and bodies, a literal combat zone: "Fire!" "Fire!" "Fire!" As the losing team pleas for more "ammunition" from the teachers, flying red balls come in contact with cadets' shoulders, knees, and backs. The loud slap of a rubber ball hitting a cadet results in another cadet crying out triumphantly: "You're dead!"

For cadets, the MEI is a literal battle zone, not of skills or knowledge but of survival. The daily practices and rituals at the MEI are saturated with militarized meanings shaping the way cadets understand their social world and themselves as they move from school ground to battleground. The militarization of public schools and educational policies could not have occurred without the simultaneous advancement of the global and domestic neoliberal political and cultural project. In this chapter, I revisit central themes in the analysis of militarization and neoliberalism and recap educational theories of socialization and resistance using the MEI and its cadets as an illustrative example. The book concludes with a discussion of current activism taking place around the demilitarization of schools and youth and proposes ways in which to challenge militarization and work to build democratic futures for youth.

Building and Sustaining Empire

Neoliberalism is a powerful pedagogy that has shaped the US society over the past 30 years in a myriad of ways as it spread the logic of free market and free enterprise. The US military ensures the success of the US financial and corporate interests and relies on a voluntary military force to keep the ranks filled. The US military is in constant need of new recruits to maintain the presence of military forces worldwide, to preserve US hegemony, and to protect the financial interests of the United States and its Western allies. While much of the discussion on neoliberalism has focused on economics and political rhetoric, it has expanded and invaded the US public education system. To ensure a continual supply of new recruits, the US military targets poor and working-class youth of color for recruitment through militarized educational programs and policies such as the DREAM Act, TTT, GI Bill, the JROTC program, DoD STARBASE and STARBASE 2.0, zero-tolerance policies, as well as militarized charter schools such as the MEI.

Additionally, the US public education system, much like US society, has been heavily influenced by capitalistic market values. Schools were historically structured to build national identity, prepare workers to occupy positions in the occupational structure of society, and as a pathway to the promise of social mobility. However, in the past 30 years, especially since the release of the 1983 report *A Nation at Risk*, which detailed the mediocrity of US public schools, neoliberal ideals of competition, accountability, choice, and efficiency became increasingly influential. Neoliberalism in the US public education system allows for and justifies educational reforms and policies such as charter schools, standardized testing, NCLB, and RTT. Each of these educational reforms was implemented as a solution to a failing US education system that overwhelmingly serves poor and working-class communities and students of color.

Despite the dismal record of neoliberal educational reforms (see Saltman 2012) and that such reforms drain resources from already underfunded public schools, choice and accountability remain powerful ideals that are widely supported by the US public as they are in line with US cultural ideals and values of competition and meritocracy. The combination of neoliberalism with the notion of a failing US public education system creates an opportunity, an emergency of crisis (Harvey 2005; Klein 2007; Saltman 2007) for alternative educational structures to emerge. The increasing militarization within public schools is one such alternative approach to education. Although the US military has had a long relationship with public schools, stretching almost 100 years, it has only been since the early 1990s that the relationship between the two flourished.

However and importantly, the militarization of US public education is mostly promoted and implemented within underfunded schools in poor communities with a high proportion of students of color. There are several reasons for the concentration of militarization within underfunded schools in poor communities that serve primarily youth of color. First, the militarization of public education coincides with the increasing militarization in US society generally (Brown 2003). Second, youth of color are constructed as violent, underachievers who do not value education, and are in desperate need of militarized discipline to set them straight (Brown 2003; Giroux 2009; Saltman 2007). Discipline, specifically militarized discipline, is seen as a solution to the menace of unrestrained, poor, and working-class youth of color. Lastly, the US military depends on volunteers to fill the ranks and relies heavily upon people of color who are victims of racism, classism, structural and educational inequality, and see the military as a "way out." Public schools and disenfranchised US youth are successful pools of recruitment for a military that is overextended. The socialization of cadets at the MEI into a culture of militarism and war increases the chances that the cadets accept the military as not only as an expected rite of passage but beneficial to their life chances, as well as increases the chances that MEI students will be open to military recruitment. In addition, military recruiters, personnel, and well-intentioned teachers sell the US military—the largest employer of youth (Kleykamp 2007)—as a viable occupational option for underprivileged youth. Many of these youth, then, view enlistment in the armed forces as a "wise career option" and a "way out" of poverty-stricken communities.

Unfortunately, choosing to enlist in the military does not guarantee transferable job skills or increased earnings. People of color are most likely to end up in combat positions, the most dangerous and the least likely units to receive training that can transfer into the civilian job market. At-risk students may view, or come to view, the military as a smart career option, but a career option among very few available options. The MEI as a militarized school needs to be understood as another form of recruitment of vulnerable youth. The military's promises of money for college and a career are empty as 50 percent of frontline troops, which are unlikely to receive technical training, are people of color (Brown 2003, 134; Levy 1998). Additionally, nonveterans have a higher rate of college degree competition (29 versus 21 percent) (Klomm Analysis Group 2000) and are more likely to complete a bachelor degree (Barry 2013; Department of Veterans Affairs 2011). Additionally, veterans are more likely to be unemployed (Bureau of Labor Statistics 2009; Humensky et al. 2013; White House 2012) and end up homeless (Bender 2009; Rourke

2007). Thus, while poor and working-class youth of color are pushed toward enlistment because of poverty, structural racism, and the promise of career and educational benefits, these promises are often unfulfilled. This is no truer than in the community of Eastmoore and the MEI.

Eastmoore is a solidly working-class community of color, and the Eastmoore School District faces many of the challenges other poor schools face, such as inadequate funding, overcrowded conditions, and violence. As outlined in chapter 3, the MEI and the Eastmoore School District successfully took advantage, through neoliberal militarized logic, the structural gaps and inequalities in the local school district to open a militarized charter school. The racial tension, presence of drug and gang activity, and academically failing and overcrowded local schools in Eastmoore pushed parents to look for educational alternatives. This coincides with parents and teachers' perception that the militarized discipline-based structure of the MEI will keep their children safe from violence and ensure academically successful futures. At the same time the MEI promotes itself as an high-achieving academic institution. While the MEI's standardized test scores are not outstanding, they are well above the average in the local school district, marking the MEI as an academically successful school in the community. This combined with the fact that the MEI has limited student enrollment and militarized uniforms that confer social status results in students, parents, and the local community perceiving the MEI as an elite institution similar to that of a private school. Thus, structural factors push parents and students away from local schools and draw them toward the MEI, increasing the enrollment at the MEI and making it a successful militarized charter school. The fact that the alternative to violent and failing local schools is a militarized charter school was not a primary factor in the cadets and parents' decision to attend the MEI, but rather, as I argue, it is a consequence of social inequality and inequitable education system that limits the choices and participation of marginalized groups.

While these structural conditions and characteristics of the Eastmoore community and the Eastmoore School District are responsible for the enrollment success of the MEI as a militarized charter school, it is important to remember that militarized charter schools and programs are particular to areas of poor or working-class and serve a high proportion of youth of color. Such schools and programs are justified through the neoliberal and militarized discourse of discipline. These are structural conditions that middle-class, suburban white students and their families do not face. Research has shown that charter schools locate in areas with a high proportion of students of color (Henig and MacDonald 2002; Gulosino and d'Entremont 2011; Manno et al. 1999; Nelson et al.

2000) and levels of poverty (CREDO 2013) as well as increase race and class segregation (Garcia 2008; Miron and Urschel 2010; Morest 2002). Thus, the MEI is an example of how concerns over struggling schools and a legacy of social, historical, and educational inequality lead to communities, parents, and students to accept militarized education as a solution – a *choice*. Neoliberal educational reforms, such a militarized charter school like the MEI, are viewed as a way out of racist, classist, and inequitable educational contexts and facilitate the Gramscian notion of consent.

Consent and Resistance to Militarization

Global capitalism requires militarization (Cabezas et al. 2007) to ensure expanding markets and the gross accumulation of wealth by neoliberal elites. In turn, the poor serve as an endless supply of desperate militarized citizens pushed into working for poverty wages or enlisting in military units. In order to maintain the violently inequitable system of militarized and economic domination, the idea of war must be normalized and justified through political maneuvering (Foucault and Ewald 2003). Foucault argues that inequality and domination justify war, but does not put an end to or alleviate domination or inequality. War, therefore, is never-ending and normalized through politics—it is war of a different form.

In the United States, the cultural and political discourse of freedom is the Trojan horse within which neoliberalism and militarization violently invade and take over our daily lives. Freedom is so deeply rooted in our cultural and political traditions that it has become common sense. Gramsci (1971) argues that common sense is the basis for the production and reproduction of class hegemony as "common sense" notions of a society, such as freedom, are in agreement with the interests of the ruling class and result in "consented coercion." That is, the interests of the ruling class are accepted in society as being interests of everyone through a network of ideas, institutions, and social relations. Thus, a militarized school is viewed as a solution to the failing, crowded, and violent Eastmoore School District. But what are the consequences of consent? What are the costs of a militarized charter school for the students who attend the MEI?

As outlined in chapter 1, social reproduction theory posits that schools serve the interests of the ruling class by reproducing and legitimating the inequitable structures in society. Deterministic models of reproduction theory (Althusser 1969; Bowles and Gintis 1976) utilize Marxist theory to argue that the structural requirements of the capitalist economic system are perpetuated through schools, and in turn, individuals fulfill predetermined class-based roles. Cultural models of social reproduction

(Bernstein 1977; Bourdieu and Passeron 1977; Heath 1983), on the other hand, argue that cultural processes within schools work to perpetuate social inequality. Both cultural and deterministic models of social reproduction overlook power and resistance within schools, and neither theoretical model provides students agency to operate within unjust and inequitable educational structures and cultures. Reproduction theorists reproduce the status quo by failing to adequately contextualize power, domination, and agency.

As a result of the shortcomings of reproduction theory, resistance theory emerged. Resistance theory examines the structural constraints, and domination individuals and groups face while simultaneously highlighting the response to these structures in the form of practices and attitudes, which are cultural as well as political. For example, critical ethnographic studies illustrate how resistance to school structures and cultural processes is an important part of school life (Everhart 1983; Fine 1991; McRobbie 1978). Resistance theory is not without its failings, however, as it often glosses over the fact that oppositional culture in schools is a reaction against historical forms of domination that exist beyond schools. Additionally, resistance theorists tend to focus on overt acts of resistance and dismiss more subtle acts that do not disempower the students by shutting them out of the skills and knowledge schools do offer that can fuel social resistance and importantly social change. To that end, this book is an extension and manipulation of resistance theory. I argue that the cultural and everyday processes at the MEI form a militarized pedagogy that socializes students to understand the world around them in militarized terms, including shaping the cadets' understanding of their futures, gender and sexual identities. Students at the MEI move from viewing themselves as students to identifying as militarized cadets.

As argued in chapter 4, the discursive use of symbols (school mascot, cannon, flag ceremony) and language (school motto, hierarchical structure of the school, school fieldtrips) utilized at the MEI shaped cadets' subjective understanding of the social world. This results in a militarized pedagogy at the MEI that perpetuates the rationality of militarization as well as the militarization of students and their identities. The militarization of the MEI leads cadets to conflate education and the military. Cadets view enlistment in the armed forces as an equally beneficial choice as going to college. However, cadets also challenge and resist the militarized practices and understandings of the MEI through their own unique gendered and queer practices and identities. As I argued in chapter 6, the MEI is a militarized space, but it is also a queerly contested space.

Chapter 5 examined the construction of gender, particularly masculinity at the MEI. Hegemonic masculinity is exemplified through violence and the warrior hero archetype. Not all cadets at the school have access to or can capitalize upon the advantages of this particular form of hegemonic masculinity, specifically girls and Black boys, but it is a powerful force that shapes social interactions, social patterns, and social identities for both boys and girls at the MEI. However, there was also resistance to the hegemonic militarized masculinity at the MEI. The "girlie-girls" actively confronted the militarized and masculine culture at the school through the embracement of traditional feminine identities by purposefully making mistakes during military drill, slowdown of physical training, and through the exhibition of improper uniforms.

Additionally, militarization of the MEI enforces heteronormative gendered and sexualized processes, identities, and behaviors through rituals and sexual scripts. Cadets at the MEI were active participants in the production of such norms as well as active participants to resist and transgress such structures and practices. Particularly in chapter 6, students resisted the gender and heteronormative practices at the school through same-sex practices such as quasi-public displays of affection, chasing games, stalking, and openly asking out other cadets of the same sex. Interestingly, the military ball, while traditionally a heteronormative space, was transformed into a contested space through the ability of students to attend as couples, in groups, or alone and the lack of gendered formal attire. Additionally, the dancing patterns of the cadets also allowed for trangression as they overwhelmingly danced with each other in large groups as well as a large number of girls dancing with other girls in sexually charged ways. Finally, a group of boisterous students actively resisted the homophobic and heteronormative militarized masculine culture of the school through the enactment of queer identities through play, laughter, and "talking shit."

However, as domination and power are never total or complete, students also actively resist, in a variety of ways, the militarized structures, processes, and culture at the MEI, particularly through their raced, gendered, and sexual bodies. This is important on two fronts. First, the resistance to militarized practices and structures at the MEI through sexuality and gender challenges the essential notions of sexuality and gender that institutions endeavor to teach us. Second, this illustrates the ability for all of us, even youth, to challenge not only militarization but also such essentialist notions of gender and sexuality and to deviate, defy, and disrupt.

While the conclusions drawn from this qualitative study cannot be generalized back to a larger population or to the US society generally, it is an important contribution to understanding how large social processes (such as neoliberalism and militarization) and public policies affect everyday practices of vulnerable youth and citizens. This book illuminates how militarization and neoliberalism operate within public education and their affect on socialization, militarization, and resistance of and by students. The MEI is an illustrative example of the nexus between increased militarization and the vast influence of neoliberal ideals in US society. This book highlights how social structures such as lack of quality education and race and class inequality create social space for the emergence and growth of militarized public education. This book is unique in that the student population at the MEI is predominately Latino, and the school is fully militarized. This book adds to the growing number of qualitative studies at public schools, but is unique in that it is a study of a charter school that moves beyond examining student enrollment or achievement outcomes toward examining how the military culture and practices influence and impact the cadets and formation of militarized, gendered, and sexual identities as well as how students actively challenge and resist heteronormative, gendered, and militarized forces.

In particular, the following arguments were made in this book. First, neoliberal ideals of choice, accountability, and efficiency are the catalyst for new types of educational forms that increase the general advancement of militarization within school structures and curriculum. Militarized schools such as the MEI are a new and flourishing trend in public education. Second, structural inequalities in the local school district and community push parents and students toward the MEI, while discipline and uniforms pull in student enrollment. Third, the militarized structure at the MEI socializes students into a culture of militarism and war. This results in students viewing themselves as cadets and the military as an equally beneficial choice as attending college. Fourth, militarization shapes the way gender is constructed at the school. Hegemonic masculinity at the school is exemplified through the condonement of violence and the warrior hero archetype. Physical violence is open to White boys, but is severely sanctioned for boys of color and girls who attempt to capitalize on the advantages of militarized hegemonic masculinity. The warrior hero archetype, on the other hand, is much more loosely constructed and is open to all students, including girls, at the MEI although in slightly varying forms. Fifth, militarization of the MEI enforces heteronormative gendered and sexualized processes and behaviors such as sexual scripts for flirting and dating as well as homophobia and homophobic slurs. Finally, a small group of feminine, self-identified "girlie-girls" actively resisted the construction of hegemonic and militarized masculinity, as well as a few students who actively challenged the heteronormative and gendered practices of the MEI. This is particularly insightful as institutions such as schools are agents of socialization, and militarized institutions serve to inculcate youth into a military understanding of their communities, lives, and bodies. However, in the face of pervasive heteromasculine and heteronormative and militarizing forces (including positive sanctions for conformity and negative sanctions for resistance), a significant number of students resisted and transgress such norms in a variety of ways.

Militarized schools such as the MEI appeal to students, parents, and community leaders, and the increasing number of such schools across the United States exemplifies the draw. However, these schools are more illustrative of the inequitable education system as marginalized individuals and communities search for an alternative to racist and classist school districts even if that alternative is militarized. Youth of color also turn to military enlistment when they view their life chances in terms of employment and ability to pay for college as bleak. Unfortunately, race and class delineate those who are the most likely to view their futures as limited. These are obstacles that wealthier, White students and communities do not confront. Curbing militarization among poor and working-class youth of color is one way to hold open the futures for these kids. Students who are directly in the line of fire of militarization need to know that one does not need to join the military to gain skills, have adventure, or acquire money for college. Halting the march of militarization in public education and returning the schoolhouse to a democratic and empowering space is one way to reduce structural inequality in education and level the unequal opportunities of the most vulnerable US citizens—the youth. How can this be achieved?

Building and Sustaining Democratic Futures/Possibilities

To build and sustain democratic futures, we must challenge both neoliberalism and militarization. First, we need to fundamentally change the way freedom is understood. Neoliberalism has taken over and occupied the meaning of freedom, reducing it to the most shallow and soulless understandings: free markets, free enterprise, individualism, and the freedom (of some) to pursue endless profit and accumulation of wealth at the expense of others. We must disentangle freedom from the greedy grip of neoliberalism and counter the blind faith in the market to destroy the belief that freedom is achieved through market practices. Harvey (2005) contends that the public discourse of freedom is so "impoverished" in the United States that alternative forms or concepts of freedom are never even seriously considered or discussed. This results in a dearth of imaginative possibilities and probabilities of what a truly democratic society could look and feel like and more importantly obscures the true libratory power of a democratic society.

Second, as neoliberalism is a direct assault on democratic principles, not only in the way in which freedom is defined, understood, and practiced but also in the exploitation of social inequality. For example, the neoliberal welfare reforms in 1996 implemented deep cuts in social expenditure and the underfunding of social programs at the same time as global shifts in labor reduced the quality of available jobs. This occurred in conjunction with rising military budgets in the United States (see Matthews 2012 for longitudinal military expenditure data). Neoliberalism and the endless pursuit of profit constructs social inequality and injustice as just another opportunity for the accumulation of capital though the privatization and corporate ransacking of public goods.

Even those citizens who cannot or do not consume (such as those victimized by historical social processes of racism, sexism, homophobia, or inequalities of class) are still profitable as neoliberalism shuttles them through the surveillance and prison industries, inequitable-yet-profitable education institutions, toward low-wage exploitative jobs and military enlistment. Neoliberalism capitalizes on inequality and militarized marginalized people by pushing them to join the military in order to meet their basic needs such as access to health care, food, shelter, and education. Thus, in order to challenge neoliberalism and militarization, we need to work against social and structural inequality. We need to ensure that the basic needs of all people are met in order to counter not only the coercive forces of neoliberalism, but militarization as well. It is estimated that in 2011 it would take $66 billion to raise every person worldwide out of absolute poverty (Chandry and Gertz 2011), a smaller amount compared to the annual $1,756 billion spent globally on military expenditure (SIPRI 2013a).

Individuals most vulnerable to neoliberalism and militarization are women, children, the poor, and people of color. Antiracist feminist analysis is useful in understanding militarization and its intersections with racism, sexism, colonialism, imperialism, and globalization. Employing an intersectional analysis of militarization offers valuable insights into how the aforementioned oppressions, such as racism and ethnocentrism, justify militarization and militarized violence (Eisenstien 2004, 2007). For example, Enloe (1989, 2000, 2007) illuminates how constructions of masculinity are infused with militarized violence. Eisenstein's (2004, 2007) feminist texts illustrate how gender and race are used to mobilize US imperialism and offers a more complex understanding of how women's rights and democratic discourses are manipulated for global hegemony. Antiracist feminist analysis also highlights how to challenge militarization. For example, Kirk (2008) uses a feminist analysis to outline four different levels in which to resist militarization: personal, community,

CONCLUSION 149

institutional, and global. This is a useful strategy for thinking about how to challenge militarization on a general level, but also how to challenge militarization within public schools. As this book has illustrated, militarization and neoliberalism permeate all aspects and levels of our lives, and our communities, so they must be resisted and challenged in a variety of ways across a multitude of analytical points. In the following section, I will outline Kirk's general approach and suggestions for resistance, respectfully adding to and broadening her insightful suggestions along the way.

Personal Level

The personal level is premised on how militarization affects us, as well as how we contribute and perpetuate militarization. Kirk (2008) lays out several different ways in which to resist militarization on the personal level; a short summary of her suggestions include choosing not to enlist in the military (I would add any militarized organization such as JROTC or the Boy Scouts of America), joining antimilitarization organizations and demonstrations, limiting financial support for the military (e.g., tax resistance), healing on a personal level from militarized violence, and demilitarizing our knowledge through critical inquiry especially in regard to mainstream media. To this list, I would first add refusal to consume militarized products and entertainment (e.g., video games, toys, films, fashion and beauty products); and second, de-militarize our everyday language by consciously refusing to use military metaphors (e.g. "The criticisms were right on *target*" or "The political *strategy* was *attacked*).

Kirk not only outlines suggestions for resistance on a personal level but also provides positive and constructive ways in which to employ our personal skills and strengths to create peaceful and nonviolent ways of living through security and sustainability. Examples include making art, building personal friendships and alliances across differences of race, ethnicity, gender, class, and sexual identity, teaching nonviolence, and participating in events that highlight nonviolence.

Community

Kirk (2008) argues that women's peace work, "where women draw on their skills and creativity to analyze their situations, define needs, and provide services" (46–47), is one of the most visible examples of resistance of militarization at the community level. Examples include establishing women's centers and workshops where women can voice their concerns about militarization and militarized violence, as well as providing access to health care and other projects that provide security and income for women affected by war and militarization. Kirk also argues that community art, media and other cultural projects can be ways in which to challenge militarization and the messages of the mainstream media, and I would add neoliberal and militarized pedagogies. It is important to learn from the students at the MEI and challenge constructions of militarized masculinity, gender, and heteronormative practices. Additionally, as racist ideologies and practices result in increased violence and militarization, we also need to maintain in our resistance repertoire a commitment to antiracist, feminist analysis and approach to community projects, workshops, and organizations. Antiracist and feminist-centered organizations that serve as models for antiracist resistance include such groups as, but not limited to, Women of Color Resource Center (UC Berkeley), Incite! Women of Color Against Violence, Women for Genuine Security, and International Women's Network against Militarism.

It is, of course, important to also build and support organizations and centers that provide spaces of peace and demilitarization for all members of a community (e.g., men, children, elderly, those marginalized across class, nation, genders, and/or sexual identities), and there are hundreds and thousands of organizations nationally and globally that are working to this end. Groups such as the Project on Youth and Non-military Opportunities, American Friends Service Committee (AFSC), Coalition Against Militarism in Schools, and the National Network Opposing Militarization of Youth (NNOMY) work to combat the militarization of youth in public schools. NNOMY maintains a database of organizations working on counterrecruitment. The database[1] can be searched by city, state, scope of organization, organization base, or particular issues of counterrecruitment. Additionally, Galaviz et al. (2011) provide an extensive list of organizations and resources for students, parents, teachers/counselors, immigrants, veterans, as well as a list of curriculum materials and videos.

Institutional

Resistance at the institutional level includes holding governments accountable to treaties and commitments such as the 2008 Child Soldiers Prevention Act and the Universal Declaration of Human Rights. We must push to change the priorities of government budgets from militarized budgets toward budgets that support social and human needs and equitably fund education. Kirk (2008) argues that for this to happen citizens must have a full understanding of the relationship between military and government budgets. Several organizations provide data on military expenditures, including the National Priorities Project, the War Resister's League, as well as the Stockholm International Peace Research Institute. Schools are one of the most important institutional sites for resistance of militarization and simultaneous fostering of democracy and democratic critical citizenry. As stated earlier, neoliberalism eyes social inequality as another avenue for profit accumulation through the privatization of public goods. There are many examples of the pillaging of the public for corporate profit (water, health care, transportation, prisons), but the violent inequality and corporate takeover of the US education system is the largest threat to democracy as schools are essential to a democratic society and provide youth with the skills and knowledge to engage and think as critical citizens. In the concluding chapter of *Capitalizing on Disaster*, Saltman (2007) argues that for a citizenry to accept inequitable and oppressive social, economic, and political institutions, they must be educated to accept these basic assumptions of domination and disproportionate power relations and this can and does often occur within schools. The learning of subjugation can take place in schools, especially militarized and corporatized schools. However, as the cadets at the MEI bravely illustrated, public schools are also sites for resistance and transformation.

As I have argued throughout this book, neoliberalism and militarization work together to dismantle the true revolutionary power of democracy. Militarized education polices, structures, and pedagogies result in uncritical, unquestioning, and obedient citizens. It is important to curb militarization of youth in order to ensure a democratic future not only for youth but also for ourselves. We must then challenge the exploitative nature of neoliberal discourse in public discussion and debate. We need to take back the original intention of public education as a democratic institution for the common good. Public schools need to return to sites of critical pedagogies, practices, and critical questioning.

Although public schools can be spaces of inculcation and repression, schools also hold true democratic and transformative power. Public schools that provide critical education provide its students with the ability to imagine and work toward a more just and equitable society. How can we ensure that the democratic promise and future of public schools is upheld? First, funding for public education must be equitable. Education needs to be released from the reliance on local property tax dollars and equitably funded on a federal level. Second, public school boards must remain publicly elected. Democratically elected school boards imclude citizens in local education decisions and are crucial to maintaining a participatory government. Finally, public funds should not be used to support private education ventures of any kind, including vouchers, tuition tax credits, for-profit charter schools or nonpublic schools.

It is also important that schools resist militarization, and there are several ways in which to accomplish this. Kirk (2008) contends that schools, as institutions, can resist militarization through curriculum changes that include alternative voices and perspectives to militarization and militarized history such as peace education as well as learning about the history of the United States and its connections with empire building though genocide, racism, sexism, and homophobia. Critical education is one way in which to counter antidemocratic forces, and public schools that critically investigate and question knowledge claims, truth, and power are an important site in which to foster critical democratic citizens and resist and challenge militarization. Kirk further argues that changes in curriculum in public schools can be ways in which to counter violence and militarization by disseminating peaceful knowledges and teaching peaceful resolutions to conflict.

Second, parents and students at public schools can demand that opt-out waivers are made easily accessible. Under NCLB, public schools must provide military recruiters with the contact information of all its pupils. Most public schools will not deny military recruiters access to the private information of its students because of the risk of losing federal funding. However, student information can be kept private and out of reach of military recruiters as schools are also required under NCLB to notify parents and students of their right to opt-out. Unfortunately, opt-out waivers have not been successful for a variety of reasons. First, opt-out forms are usually sent home with students at the beginning of the year along with other information packets, handbooks, forms, and notifications. As a result, opt-out waivers often get overlooked. Second, some schools do not have any formal policy in place regarding how or what type of form or waiver to accept. Finally, there is often a deadline for returning the opt-out waivers to the school. These deadlines are not always clear. Because of these above reasons, it is important for parents and students alike to know their school's policy regarding opt-out waivers and to push the school to implement one if it is not in place. It has also proven useful to pressure schools to provide multiple ways in which to opt out such as mailings and online forms.

Unfortunately, opting out is not enough to be certain that a child's information is kept private from military recruiters. The ASVAB is a test that is administered and controlled by the military. It was originally designed to predict academic and occupational success within the military, but it is now used as an aptitude or career placement test in high schools across the country. The results of the test along with the test takers' private information are forwarded to the military. Military recruiters use this information to identify recruitment leads who meet the basic requirements for military service. Thus, by participating in the ASVAB the military once again gains access to students' private information. Students and parents need to be aware that the ASVAB is not mandatory and that no school can force a student to participate in the test. Students should refuse to take the ASVAB and urge schools to offer alternative aptitude tests.

In addition, during the 1980s, several lawsuits by counterrecruitment organizations resulted in federal rulings that allow for equal access of counterrecruitment organizations in high schools (see, for example, San Diego Committee v. Governing Board of Grossmont Union High School District [790 F.2d 1471 (9th Cir. 1986)]). According to the ruling, once a forum has been set up on a controversial topic, such as military recruitment, access to the forum can be limited but only so long as it is not a basis or viewpoint for discrimination. Thus, according to this ruling, if one viewpoint has been allowed, alternative viewpoints must also be allowed. This legal information can be utilized to set up counterrecruitment tables or distribute antirecruitment pamphlets at a school site.

Furthermore, as a strategy against militarization of youth and schools, it is important to oppose the Joint Advertising and Market Research Studies (JAMRS). This is a database compiled by the Pentagon of the private information for individuals, such as address, email, social security number, GPA, ethnicity, and area of study, of those who are of recruitment age. The database is distributed monthly to the different branches of the armed services for recruitment purposes. The database also follows youth attitudes towards the military and uses this information to increase recruitment through targeted advertising. One can opt out of having his or her child's information included in the database by writing to the Pentagon. There is not, at this time, an electronic opt-out procedure. A JAMRS Opt-Out form is available on the New York Civil Liberties Union and NNOMY websites. By opting out, your child's information is moved to a "suppression file." The Pentagon retains the information, but does not release it.

Lastly, the JROTC must be eliminated. Almost half of all students who participate in the JROTC join the military. JROTC programs incur heavy start-up and maintenance costs to local schools. Additionally, JROTC instructors do not have to be certified instructors and are usually retired military personnel. Talk to students and parents about the costs and risks associated with the JROTC program and encourage students not to join. If enrollment falls below 100 for two years, the JROTC must leave the school. For this to occur, it is important to offer youth alternatives to militarization, JROTC programs, and enlistment. We need to offer students not only counterrecruitment information but also alternatives to the military in terms of career, educational funding, and job skills/training. The AFSC has great resources on their website (http://afsc.org/key-issues/issue/community-peace-building) that details alternatives to military enlistment for youth, provides examples of careers in social justice and peace and lists ways to pay for higher education that do not include military enlistment.

Global

Confronting neoliberalism and militarization solely with the borders of the United States is not enough. Militarization and neoliberalism are both global projects; they cannot be countered with domestically based challenges and alternatives. We must work to create global movements and global alliances as well as more just and humane public spheres for youth globally. Militarization is justified in terms of security, but the security of corporate or state interests rather than the interests of the citizenry and basic human needs. One way to do this is by changing the way we understand global politics, foreign policy, and security. Sutton and Novkov (2008) argue for a "human security" paradigm in which the focus is on basic human needs rather than the needs of global or neoliberal policies of the state (surveillance, militarization, policing). Additionally, Vine (2009) argues for placing the needs of human lives at the forefront of foreign policy. He favors a "humanpolitik" approach to foreign policy, one based on "diplomacy, international cooperation 'non-aggression' and the protection of human security as the best way to protect the security of the US and, ultimately, the world" (175). Both these approaches need to take into consideration the unique positioning of global citizens and how security takes into account inequalities such as race/ethnic injustice, gender-based inequalities, poverty, ethnocentrism, homophobia, but also how nations and citizens of such nations are privileged.

One way to accomplish this is to forge global alliances and global identities that start from the understanding that we are all, in a sense, in this together, as global peace and equality results in community security and vice versa. These organizations must be formed on a dedication to creating global relationships that are equitable, respectful, and democratic. Such organization must work to prevent conflict and promote global peace initiatives. There are a variety of organizations that are working and networking globally to counter militarization and violence. Kirk (2008) discusses a number of them, but a few examples include International Women's Network against Militarism, Global Action to Prevent War and Conflict, Federation of African Women's Peace Networks, and the Women's International League for Peace and Freedom.

Reisiting militarization across global, institutional, community, and personal levels may seem impossible, especially considering neoliberal discourse that pervades the way in which freedom and democracy are understood in the United States, but I conclude by drawing on a quote from Bassichis et al.(2011), who outlined a strategy for building an abolitionist trans and queer movement, "*impossibility may be our only possibility*" (36; emphasis original).

Notes

1 Introduction

1. For example, laws that California charter schools must comply with include all state and federal constitutions; the California Charter Schools Act; all federal laws such as the Individuals with Disabilities Education Act, Americans with Disabilities Act, and Rehabilitation Act; all laws that are specifically a condition of funding for a specific program; laws establishing minimum age for school attendance; laws governing independent study; Educational Employees Relations Act; state pupil testing requirements; and specific provisions of law related to teachers' requirements and employee relations.
2. As of 2011, women made up 15 percent of the total active military (Department of Defense 2012b). This compares to less than 2 percent in 1950 (Statistical Abstract of the United States 2006).
3. Meaning that less than 5 percent of the students at that particular school are white.
4. A commandant is equal to the position of principal in nonmilitary public schools.
5. Observation consisted of simply observing daily activities at the MEI. Participant observation, however, meant that I was an active participant at the MEI filling particular social roles and duties as needed.
6. The group consisted of 3 Black students, 3 white students, and 6 Latino/a students.
7. Both the school secretary and security officer were white women. Both commandants were white men.
8. In the first year, the teaching staff consisted of one Black man and a White woman. The second year comprised two Black men, one White man, and 1 Latina. In the third year, the teaching staff consisted of two Black men, one White man, one White woman, 1 Latino, and 1 Latina.
9. Five women and one man, four of whom were White and two Latinas.
10. Two Black male teachers, a white male teacher, and two White female teachers.
11. Two of the parents were Latina and one parent was White.

2 Schools in the Crosshairs: Neoliberalism, Militarization, and Public Education

1. Possibly the most well-known example is the poster of Uncle Sam by James Montgomery Flagg during World War I that read "I want YOU for U.S. Army."
2. The Deferred Action for Childhood Arrivals (DACA) is a memorandum signed by the Obama administration in 2012. It is a discretionary grant in which young people can apply for "deferred action" if they came to the United States as children and have pursued military or education service. A person with deferred action status can apply for employment authorization and is considered legal. However, DACA does not lead to permanent residency or citizenship.
3. The following states currently allow undocumented students to receive in-state college tuition benefits: Illinois, Utah, Texas, Nebraska, Washington, New York, Kansas, Oklahoma, California, and New Mexico.

3 Sending Good Kids to Military School: Why Parents Choose the MEI?

1. Title I funds are part of the Elementary and Secondary Education Act and are intended to improve educational standards for high-poverty schools and/or schools with struggling students who are at risk for failing.
2. STAR (Standardized Testing and Reporting) tests measure performance in grades two through eleven in the state of California.
3. According to the California Department of Education website, emergency permits are for those teachers "who do not qualify for a credential or internship but meet minimum requirements."
4. It tests students in English language arts in grades two through eleven, math in grades two through seven, science in grades five, eight, and ten, and history social science in grades eight and eleven.
5. The corps of cadets refers to the student population at the MEI.

4 Reading, Writing, Arithmetic, and War: Militarized Pedagogy and Militarized Futures

1. The Middle School Cadet Corps is an after-school program for middle school students and utilizes a military structure for instruction.
2. The ASVAB test is administered and controlled by the military. It was originally designed to predict academic and occupation success within the military, but it is now used as an aptitude or career placement test in high schools across the country. The results of the tests, along with the test takers' private information, are forwarded to the military, and military recruiters use this

information to identify recruitment leads that meet the basic requirements for military service.
3. The Delayed Entry Program allows youth to enlist in the armed forces and specify a future reporting date. This program is pushed for seniors who are unsure of their plans after high school. The program allows for individuals to set their reporting date after high school graduation or eighteenth birthday, for example.
4. The cannon ceremony was temporarily stopped in the second year of MEI, as the new location does not have a flagpole or a viable area to fire the cannon. The cannon is still widely used, however, in other school activities and ceremonies such as parades. Major West plans on continuing the cannon and flag ceremony when a permanent location is found for MEI.

6 Ask, Tell, Talk Back: Queering Resistance to Gendered Heteronormativity

1. Camp (style) performance in popular culture is probably the best-known example of Atkinson and DePalma's concept of burlesquing. See Sontag (1964) for a discussion of Camp.
2. For example, Ashanti, an assertive and bright South Asian seventh grader, was very confused about pubic lice or "crabs," as she termed them.
3. While the authors argue that there are as many sexual scripts as there are individuals, there are in fact scripts that form patterns and trends at the macrosocial level that guide sexual behavior. There are three levels of scripts that work simultaneously to construct social-sexual interactions—first, cultural scenarios or norms and standards that dictate appropriate sexual behavior; second, interpersonal scenarios where individuals interpret these larger society-wide scripts into specific and individual contexts; third, intrapsychic scripts through which multiple desires and fantasies (yours and others) are mediated.
4. This as opposed to the uniforms worn usually at the MEI, which consisted of camouflage pants, long sleeve top, and cap purchased at the army supply store.

7 Conclusion

1. The database can be accessed at: http://www.nnomy.org/joomla/index.php?option=com_sobi2&Itemid=178.

References

Adelman, Madelaine. 2003. "The Military, Militarism, and the Militarization of Domestic Violence." *Violence Against Women* 9 (9): 1118–1152.
Adler, Patricia A., Steven J. Kless, and Peter Adler. 1992. "Socialization to Gender Roles: Popularity Among Elementary School Boys and Girls." *Sociology of Education* 65 (3): 169–187.
Allen, Karie. 2004. "Voters OK Bond Measure." *The Press-Enterprise*, November 4, B1.
Althusser, Louis. 1969. *For Marx*. New York, NY: Vintage Books.
———. 1971. "Ideology and the Ideological State Apparatuses." In *Lenin and Philosophy and Other Essays*, translated by Ben Brewster. New York, NY: Monthly ReviewPress.
American Association of University Women Educational Foundation. 1992. *How Schools Shortchange Girls: The AAUW Report: A Study of Major Findings on Girls in Education*. New York: Marlowe and Company.
ACLU (American Civil Liberties Union). 2008. *Soldiers of Misfortune: Abusive U.S. Military Recruitment and Failure to Protect Child Soldiers*. New York: American Civil Liberties Union. http://www.aclu.org/intlhumanrights/gen/35245pub20080513.html.
Amrhein, Saundra. 2009. "College Dream Drives an Army." *St. Petersburg Times (Florida)*, March 21, A1.
Amin, Samir. 2004. "U.S. Imperialism, Europe, and the Middle East." *Monthly Review* 56 (6): 13–33.
Anderson, David C. 1998. "Curriculum, Culture and Community: The Challenge of School Violence." *Crime and Justice* 24: 317–363.
Anderson, Margaret L. 1988. *Thinking about Women: Sociological Perspectives on Sex and Gender*. New York: MacMillan.
Andre-Bechely, Lois. 2007. "Finding Space and Managing Distance: Public School Choice in an Urban California District." *Urban Studies* 44 (7): 1355–1376.
Angrist, Joshua D. 1990. "Lifetime Earnings and the Vietnam Era Draft Lottery: Evidence from Social Security Administrate Records." *The American Economic Review* 80 (3): 313–336.
Apple, Michael W. 2001. "Comparing Neo-Liberal Projects and Inequality in Education." *Comparative Education* 37 (4): 409–423.
———. 2013. *Education and Power*. New York: Routledge.
Arce, Josephine, Debra Luna, Ali Borjian, and Marguerite Conrad. 2005. "No Child Left Behind: Who Wins Who Loses?" *Social Justice* 32 (3): 56–71.

Associated Press. 2006. "Charges Fuel Debate over Military Recruiters' Access to Students." April 16.
Atkinson, Elizabeth, and Renee DePalma. 2008. "Imagining the Homonormative: Performative Subversion in Education for Social Justice." *British Journal of Sociology of Education* 29 (1): 25–35.
Baily, Beth. 2007. "The Army in the Marketplace: Recruiting an All-Volunteer Force." *Journal of American History* 94 (1): 47–74.
Banchero, Stephanie. 2010a. "Daley School Plan Fails to Make Grade." *Chicago Tribune*, January 17. http://articles.chicagotribune.com/2010-01-17/news/1001160276_1_charter-schools-chicago-reform-urban-education.
———. 2010b. "Literacy Scores Stall in Inner Cities." *The Wall Street Journal*, May 20. http://online.wsj.com/news/articles/SB10001424052748703559004575256143731101672.
Banchero, Stephanie, and C. Sadovi. 2007. "Chicago Military Schools: Reading, Writing, Recruiting." *Chicago Tribune*, October 15. http://articles.chicagotribune.com/2007-10-15/news/070140434_1_junior-rotcmilitary-academies-chicago-public-schools.
Bancroft, Kim. 2009. "To Have and to Have Not: The Socioeconomics of Charter Schools." *Education and Urban Society* 41 (2): 248–279.
Barley, Stephen R. 1998. "Military Downsizing and the Career Aspects of Youth." *Annals of the American Academy of Political and Social Science* 559: 141–57.
Barry, Catherine N. 2013. "Moving on Up? U.S. Military Service, Education and Labor Market Mobility among Children of Immigrants." PhD dissertation, University of California, Berkeley.
Basham, Victoria M. 2011. "Kids with Guns: Militarization, Masculinities, Moral Panic, and (Dis)Organized Violence." In *The Militarization of Childhood*, edited by Beier J. Marshall, 175–193. New York: Palgrave Macmillan.
Bassichis, Morgan, Alexander Lee, and Spade Dean. 2011. "Building an Abolitionist Trans and Queer Movement with Everything We've Got." In *Captive Genders: Trans Embodiment and the Prison Industrial Complex*, edited by Nat Smith and Eric A. Stanley, 15–40. Oakland, CA: AK Press.
Basu, Moni. 2013. "Why Suicide Rate among Veterans May be More than 22 a Day." *CNN*, November 14. http://www.cnn.com/2013/09/21/us/22-veteransuicides-a-day/..
Bakhtin, Mikhail. 1984 [1936]. *Rabelais and His World*. Bloomington: Indiana University Press.
Belkin, Aaron. 2001. "The Pentagon's Gay Ban Is Not Based on Military Necessity." *Journal of Homosexuality* 14 (1): 103–119.
Bender, Bryan. 2009. "More Female Veterans Are Winding up Homeless." *The Boston Globe*, July 6, A1.
Berends, Mark, Caroline Watral, Bettie Teasley, and Anna Nicotera. 2008. "Charter School Effects on Achievement: Where We Are and Where We're Going." In *Charter School Outcomes*, edited by Mark Berends, Matthew G. Springer, Herbert J. Walberg, 243–266. New York: Lawrence Erlbaum Associates.

Berlowitz, Marvin J., and Nathan A. Long. 2003. "The Proliferation of the JROTC: Educational Reform or Militarization." In *Education as Enforcement: The Militarization and Corporatization of Schools*, edited by Kenneth J. Saltman and David A. Gabbard, 163–174. New York: RoutledgeFalmer.

Bernstein, Basil. 1977. "Social Class, Language, and Socialization." In *Power and Ideology in Education*, edited by Jerome Karabel and A. H. Halsey, 473–486. New York: Oxford University Press.

Best, Raphaela. 1983. *We All Have Scars*. Bloomington: Indiana University Press.

Bettie, Julie. 2003. *Women without Class: Girls, Race, and Identity*. Berkeley: University of California Press.

Biggs, Carl Leon. 2010. "Junior Reserve Officers Training Corps: A Comparison of Achievement, High School Graduation, College Enrollment, and Military Enlistment Rates of High School Students in Missouri." EdD diss., University of Arkansas.

Blackburn, Mollie V. 2004. "Agency in Borderland Discourses: Engaging in Gaybonics for Pleasure, Subversion, and Retaliation." In *Youth and Sexuality: Pleasure, Subversion, and Insubordination In and Out of Schools*, edited by Mary Louise Rasmussen, Eric Roles, and Susan Talburt, 177–199. New York: Palgrave Macmillan.

Blanc, A. 2001. "The Effect of Power in Sexual Relationships on Reproductive and Sexual Health: An Examination of the Evidence." *Studies in Family Planning* 32 (3): 189–213.

Boje, David. 2003. "ROTC at New Mexico State University Targets Latino Students, Giving Confidential Data to Recruiters." www.PeaceAware.com.

Bourdieu, Pierre. 1977. *Outline of Theory and Practice*. Cambridge: Cambridge University Press.

———. 1979. "Symbolic Violence." *Critique of Anthropology* 4 (3/14): 77–85.

Bourdieu, Pierre, and J. C. Passeron. 1977. *Reproduction in Education, Society and Culture*. London: Sage.

Bowles, Samuel, and Herbert Gintis. 1976. *Schooling in Capitalist America: Educational Reform and the Contradictions of Economic Life*. New York: Basic Books.

Bracey, Gerald. 2005. "No Child Left Behind: Where Does the Money Go?" Tempe, AZ: Education Policy Studies Laboratory. http://files.eric.ed.gov/fulltext/ED508523.pdf.

Brandon, Mark E. 2003. "War and American Constitutional Order." *Vanderbilt Law Review* 56 (6): 1815–1869.

Brown, Enora R. 2011. "Freedom for Some, Discipline for 'Others: The Structure of Inequality in Education.'" In *Education as Enforcement: The Militarization and Corporatization of Schools*, 2nd ed, edited by Kenneth J. Saltman and David A. Gabbard, 130–164. New York: RoutledgeFalmer.

Brown, Liz, and Eric Gutstein. 2009. "The Charter Difference: A Comparison of Chicago Charter and Neighborhood High Schools." A Collaborative for Equity and Justice in Education Report University of Illinois-Chicago

College of Education. http://ceje.uic.edu/wpcontent/uploads/2013/11/CharterDifference.pdf
Brown, Lyn Mikel. 2003. *Girlfighting: Betrayal and Rejection among Girls*. New York: New York University Press.
Brown, Wendy. 2005. *Edgework*. Princeton, NJ: Princeton University Press.
Brownstein, Ronald. 2003. "Implementing No Child Left Behind." In *The Future of School Choice*, edited by Paul E. Peterson, 213–226. Stanford, CA: Hoover Institutional Press Publication.
Bryant, Richard R., V. A. Samaranayake, and Allen Wilhite. 1993. "The Effect of Military Service on the Subsequent Civilian Wage of the Post-Vietnam Veteran," *Quarterly Review of Economics and Finance* 33 (1): 15–31.
Buck, Howard. 2004. "Parents Tell Military not to Call their Children." *The Columbian*, October 21, A1.
Bureau of Labor Statistics. 2009. *Employment Situation of Veterans: 2008*. Washington, DC: US Department of Labor.
Burn, Shawn Meghan. 2000. "Heterosexuals' Use of 'Fag' and 'Queer' to Deride One Another: A Contribution to Heterosexism and Stigma." *Journal of Homosexuality* 40 (2): 1–11.
Butler, Judith. 1995. "Melancholy Gender/Refused Identification." In *Constructing Masculinity*, edited by Maurice Berger, Brian Willis, and Simon Watson, 21–36. New York: Routledge.
Cabezas, Amalia L., Ellen Reese, and Marguerite Waller. 2007. "Introduction." In *The Wages of Empire,* edited by Amalia L. Cabezas, Ellen Reese, and Marguerite Waller, 1–15. Boulder, CO: Paradigm Publishers.
California Department of Education. 2009. *Number of English Learners by Language: School Report*. http://dq.cde.ca.gov/dataquest/LEP.
Chandy, Laurence, and Geoffrey Gertz. 2011. *Poverty in Numbers: The Changing State of Global Poverty from 2005 to 2015*. Washington, DC: Brookings Institution.
Chery, Carl. 2003. "U.S. Army Targets Black Hip-Hop Fans." *Wire/Daily Hip-Hop News*, October 21. www.sohh.com/ariticle_print.php?content_ID=5162.
Children's Defense Fund. 1975. *School Suspensions: Are They Helping Children?* Cambridge, MA: Washington Research Project.
Chubb, John E., and Terry M. Moe. 1990. *Politics, Markets and America's Schools*. Washington, DC: Brookings Institution.
Clark, Philip. 2004. *Trading Books for Soldiers: The True Cost of JROTC*. Philadelphia, PA: American Friends Service Committee. www.afsc.org/youthmil/jrotc/jrotcost.htm.
Cobb, C. D., and G. V. Glass. 1999. "Ethnic Segregation in Arizona Charter Schools." *Education Policy Analysis Archives* 7 (1): 1. http://epaa.asu.edu/epaa/v7n1/.
Coleman, James S., E. Q. Campbell, C. J. Hobson, J. McPartland, A. M. Mood, F. D. Weinfeld, and R. L. York. 1966. *Equality of Education Opportunity*. Washington, DC: US Government Printing Office.
Connell, Bob. 1992. "Masculinity, Violence and War." In *Men's Lives*, edited by Michael S. Kimmel and Michael A. Messner, 125–130. Boston, MA: Allyn and Bacon.

Connell, R. W. 1987. *Gender and Power: Society, the Person and Sexual Politics.* Stanford: Stanford University Press.
——. 1995. *Masculinities.* Los Angeles: University of California Press.
——. 1996. "Teaching the Boys: New Research on Masculinity, and Gender Strategies for Schools." *Teacher's College Record* 98 (2): 206–235.
——. 2000. *The Men and the Boys.* Los Angeles: University of California Press.
Coumbe, Arthur T., and Lee S. Harford. 1996. *U.S. Army Cadet Command: The 10 Year History.* Fort Monroe, VA: Office of the Command Historian, US Army.
Cowen Institute. 2013. *The State of Public Education in New Orleans.* New Orleans: LA: Tulane University, Scott S. Cowen Institute for Public Education Initiatives.
Craik, Jennifer. 2005. *Uniforms Exposed (Dress, Body, Culture).* Oxford: Berg Publishers.
CREDO. 2013. *National Charter School Study Executive Summary.* Standford, CA: Stanford University, Center for Research on Education Outcomes.
Crouse, James, and Dale Trusheim. 1988. *The Case against the SAT.* Chicago, IL: University of Chicago Press.
Dalley, Phyllis, and Mark David Campbell. 2003. "Constructing and Contesting Discourses of Heteronormativity: An Ethnographic Study of Youth in a Francophone High School in Canada." *Journal of Language, Identity, and Education* 5 (1): 11–29.
Darling-Hammond, Linda. 2004. "From 'Separate but Equal' to 'No Child Left Behind': The Collusion of New Standards and Old Inequalities." In *Many Children Left Behind: How the No Child Left Behind Act Is Damaging our Children and our Schools,* edited by Deborah Meier and George Wood, 3–32. Boston, MA: Beacon Press.
——. 2010. *The Flat World and Education.* New York: Teachers College Press.
Darling-Hammond, Linda, and Peter Youngs. 2002. "Defining 'Highly Qualified Teachers': What Does 'Scientifically-Based Research' Actually Tell Us?" *Educational Researcher* 31 (9): 13–25.
Dawson, Graham. 1994. *Soldier Heroes, British Adventure, Empire and the Imagining of Masculinity.* London: Routledge.
Department of Defense. 2000. *Population Representation in the Military Services Fiscal Year 1999.* Washington, DC: Office of the Assistant Secretary of Defense.
——. 2011. *Department of Defense: Youth Poll: Overview Report.* Alexandria, VA: JAMRS. www.jamrs.defense.gov.
——. 2012a. *Base Structure Report: Fiscal Year 2012 Baseline.* Washington, DC: Office of the Deputy Under Secretary of Defense (Installations and Environment).
——. 2012b. *2012 Demographics: Profile of the Military Community.* Washington, DC: Department of Veteran's Affairs. Office of the Deputy Under Secretary of Defense.
Department of Veteran's Affairs 2005. *GI-BILL History.* http://www.gibill.va.gov/GI_Bill_Info/history.htm.

REFERENCES

Department of Veteran's Affairs 2011. *Educational Attainment of Veterans: 2000 to 2009*. Washington, DC: National Center for Beterans Analysis and Statistics.

DeVoe, Jill F., Katharin Peter, Margaret Noonan, Thomas D. Snyder, and Katrina Baum. 2005. *Indicators of School Crime and Safety 2005*. Washington, DC: US Departments of Education and Justice.

Eder, Donna. 2003. *School Talk*. New Brunswick: Rutgers University Press.

Eder, Donna, and Maureen T. Hallinan. 1978. "Sex Differences in Children's Friendships." *American Sociological Review* 43: 237–250.

Eder, Donna, and Stephen Parker. 1987. "The Cultural Production and Reproduction of Gender: The Effect of Extracurricular Activities on Peer-group Culture." *Sociology of Education* 60 (3): 200–13.

Eighmey, John. 2006. "Why Do Youth Enlist? Identification of Underlying Themes." *Armed Forces and Society* 32 (2): 307–328.

Eisenstein, Zillah. 2004. *Against Empire: Feminisms, Racism and the West*. London: Zed Books.

———. 2007. *Sexual Decoys: Gender, Race, and War in Imperial Democracy*. New York: Zed Books.

Elliott, Stuart. 2013. "Army Tries a Reality Style for Recruitment." *The New York Times*, May 22, B8.

Emmanuel, Adeshina. 2012. "Like the Army Itself, Recruiters Prepare to Make Do with Less." *The New York Times*. September 5. http://www.nytimes.com/2012/09/06/us/like-the-army-itself-recruiters-prepare-to-make-do-with-less.html.

Enloe, Cynthia. 1989. *Bananas, Beaches and Bases: Making Feminist Sense of International Politics*. Los Angeles: University of California Press.

———. 2000. *Maneuvers: The International Politics of Militarizing Women's Lives*. Los Angeles: University of California Press.

———. 2007. *Globalization and Militarism: Feminists Make the Link*. New York: Rowman & Littlefield Publishers, Inc.

Epps, Edgar T. 2002. "Race, Class and Educational Opportunity: Trends in the Sociology of Education." In *2001 Race Odyssey: African Americans and Sociology*, edited by Bruce R. Hare, 164–177. Syracuse, NY: Syracuse University Press.

Epstein, D., and R. Johnson. 1998. *Schooling Sexualities*. Buckingham: Open University Press.

Esterberg, Kristin G. 1996. "'A Certain Swagger When I Walk': Performing Lesbian Identity." In *Queer Theory/Sociology*, edited by Steven Seidman, 259–279. Cambridge, MA: Blackwell Publishers, Inc.

Everhart, Robert B. 1983. *Reading, Writing, and Resistance: Adolescence and Labor in a Junior High School*. Boston, MA: Routledge and Kegan Paul.

Expect More. 2006. "Detailed Information on the Junior Reserve Officer Training Corps Assessment." http://www.whitehouse.gov/sites/default/files/omb/assets/omb/expectmore/detail/ 10003233.2006.html.

Fact Sheet: The Race to the Top. 2009. Office of the Press Secretary, The White House. http://www.whitehouse.gov/the-press-office/fact-sheet-race-top.

Fausto-Sterling, Ann. 1995. "How to Build a Man." In *Constructing Masculinity*, edited by Maurice Berger, Brian Wallis, and Simon Watson, 127–134. New York: Routledge.
Feistritzer, C. E. 2005. "Profile of Troops to Teachers." Washington, DC: National Center for Education Information. http://www.ncei.com/surveys.html.
Ferguson, Ann. 2000. *Bad Boys: Public Schools in the Making of Black Masculinity*. Ann Arbor: University of Michigan Press.
Fine, Michelle. 1991. *Framing Dropouts: Notes on the Politics of an Urban Public High School*. Albany, NY: State University of New York Press.
Fleming, Jacqueline, and Nancy Garcia. 1998. "Are Standardized Tests Fair to African Americans? Predictive Validity of the SAT in Black and White Institutions." *Journal of Higher Education* 69 (5): 471–495.
Flores-Gonzalez, Nilda. 2005. "Popularity versus Respect: School Structure, Peer Groups and Latino Academic Achievement." *International Journal of Qualitative Studies in Education* 18 (5): 625–642.
Foucault, Michel. 1972. *The Archaeology of Knowledge and the Discourse on Language*. New York: Pantheon.
———. 1975. *Discipline and Punish: The Birth of the Prison*. New York: Vintage Books.
———. 1977. *Discipline and Punish: The Birth of the Prison*. New York: Random House.
———. 1978. *The History of Sexuality, Vol. 1: An Introduction*. New York: Pantheon.
———. 1990. *The History of Sexuality*. New York: Random House.
Foucault, Michel, and François Ewald. 2003. *Society Must be Defended: Lectures at the Collège de France, 1975–1976*, vol. 3. New York: Macmillan.
Foucault, Michel, and Jay Miskoweic. 1986. "Of Other Spaces." *Diacritics* 16 (1): 22–27.
Fordham, Signithia, and John U. Ogbu. 1986. "Black Students, School Success: Coping with the 'Burden of Acting White'." *Urban Review* 18 (3): 176–206.
Francis, B., and Christie Skelton. 2001. "Men Teachers and the Construction of Heterosexual Masculinity in the Classroom." *Sex Education* 1 (1): 9–21.
Fraser, Nancy. 1990. "Rethinking the Public Sphere: A Contribution to the Critique of Actually Existing Democracy." *Social Text* 25/26: 56–80.
Friedman, Milton. 1955. *The Role of Government in Education*. New Brunswick, NJ: Rutgers University Press.
———. 1962. *Capitalism and Freedom*. Chicago: University of Chicago Press.
Fry, Rick. 2007. *The Changing Racial and Ethnic Composition of U.S. Public Schools*. Washington, DC: Pew Hispanic Center. http://www.pewhispanic.org/2007/08/30/the-changing-racial-and-ethnic- composition-of-us-public-schools/.
Fusarelli, Lance D. 2002. "Texas: Charter Schools and the Struggle for Equity." In *The Charter School Landscape*, edited by Sandra Vergari, 175–191. Pittsburgh, PA: University of Pittsburgh Press.
Gagnon, John H. 1990. "The Explicit and Implicit Use of the Scripting Perspective in Sex Research." *Annual Review of Sex Research* 1 (1): 1–43.

Gagnon, John H., and William Simon. 2005 [1973]. *Sexual Conduct: The Social Sources of Human Sexuality*. London: AldineTransaction.

Galaviz, Brian, Jesus Palafox, Erica R. Meiners, and Therese Quinn. 2011. "The Militarization and the Privatization of Public Schools." *Berkeley Review of Education* 2 (1): 27–45.

Gamson, Joshua, and Dawne Moon. 2004. "The Sociology of Sexualities: Queer and Beyond." *Annual Review of Sociology* 30: 47–64.

Garcia, David R. 2008. "Academic and Racial Segregation in Charter Schools: Do Parents Sort Students Into Specialized Charter Schools?" *Education and Urban Society* 40 (5): 590–612.

Gilligan, Carol. 1982. *In a Different Voice: Psychological Theory and Women's Development*. Cambridge, MA: Harvard University Press.

Gilmore. Ruth. 2000. "Behind the Power of 41 Bullets: An Interview with Ruth Wilson Gilmore; What is Domestic Militarization and How Did it Come About?" *Colorlines*, January 31: 16–21.

Giroux, Henry A. 1983a. "Theories of Reproduction and Resistance in the New Sociology of Education: A Critical Analysis." *Harvard Educational Review* 53 (3): 289–293.

———. 1983b. *Theory and Resistance in Education: A Pedagogy for the Opposition*. New York: Bergin and Garvey.

———. 1996. *Fugitive Cultures: Race, Violence and Youth*. New York: Routledge.

———. 2001. *Theory and Resistance in Education: Toward a Pedagogy for the Opposition*. Westport: CT: Bergin & Garvey.

———. 2003. *Public Spaces, Private Lives: Democracy Beyond 9/11*. New York: Rowman & Littlefield Publishers, Inc.

———. 2004. *The Terror of Neoliberalism: Authoritarianism and the Eclipse of Democracy*. Boulder, CO: Paradigm Publishers.

———. 2007. *The University in Chains: Confronting the Military-Industrial-Academic Complex*. Boulder, CO: Paradigm Publishers.

———. 2011. *On Critical Pedagogy*. London: Bloomsbury Publishing.

———. 2009. *Youth in a Suspect Society*. New York, NY: Palgrave.

Glaser, Barney, and Anselm Strauss. 1967. *The Discovery of Grounded Theory*. Chicago: Aldine.

Goe, Laura. 2007. *The Link Between Teacher Quality and Student Outcomes: A Research Synthesis*. Washington, DC: National Comprehensive Center for Teacher Quality.

Goldberg, Matthew S., and John T. Warner. 1987. "Military Experience, Civilian Experience, and the Earnings of Veterans." *The Journal of Human Resources* 22 (1): 62–81.

Goldhaber, Dan, Kacey Guin, Jeffrey R. Henig, Frederick M. Hess, and Janet A. Weiss. 2005. "How School Choice Affects Students Who Do Not Choose." In *Getting Choice Right*, edited by Julian R. Betts and Tom Loveless, 101–129. Washington, DC: The Brookings Institution.

Goldstein, Joshua S. 2001. *War and Gender*. Cambridge: Cambridge University Press.

Goslin, David A. 1967. *Criticism of Standardized Tests and Testing.* Washington, DC: US Department of Health, Education, and Welfare; Office of Education.
Gramsci, Antonio, Geoffrey Nowell-Smith, and Quintin Hoare. 1971. *Selections from the Prison Notebooks of Antonio Gramsci,* edited and translated by Quintin Hoare and Geoffrey Nowell Smith. New York: International Publishers.
Griffin, C., and S. Lees. 1997. "Editorial: Masculinities in Education." *Gender and Education* 9 (1): 5–8.
Grossman, Arnold H. 1997. "Growing up with a 'Spoiled Identity': Lesbian, Gay, and Bisexual Youth at Risk." *Journal of Gay and Lesbian Social Services* 6 (3): 45–56.
Gulosino, C. and C. d'Entremont. 2011. "Circles of Influence: An Analysis of Charter School Location and Racial Patterns at Varying Geographic Scales." *Education Policy Analysis Archives* 19 (8): 1–29.
Gusterson, Hugh. 2009. "Militarizing Knowledge." In *The Counter-Counterinsurgency Manual: Or, Notes on Demilitarizing American Society,* edited by Network of Concerned Anthropologists Steering Committee, 39–55. Chicago: Prickly Paradigm Press.
Halberstam, Judith. 1998. *Female Masculinity.* Durham: Duke University Press.
Hall, Stuart. 1981. "Cultural Studies: Two Paradigms." In *Culture, Ideology, and Social Process,* edited by Tony Bennett et al. London: Batsford Academic & Educational.
Harrison, Deborah. 2003. "Violence in the Military Community." In *Military Masculinities: Identity and the State,* edited by Paul R. Higate, 72–90. Westport, CT: Praeger Publishers.
Harvey, David. 2005. *A Brief History of Neoliberalism.* New York: Oxford University Press.
Haymes, Stephen Nathan. 1995. *Race, Culture, and the City.* Albany, NY: State University of New York.
Haywood, Chris. 1996. "Out of the Curriculum: Sex Talking, Talking Sex." *Curriculum Studies* 4 (2): 229–251.
Haywood, Christian, and Martin Mac an Ghaill. 1996. "Schooling Masculinities." In *Understanding Masculinities: Social Relations and Cultural Arenas,* edited by Martin Mac an Ghaill, 50–60. Buckingham: Open University Press.
Heath, Shirley Brice. 1983. *Ways with Words.* Cambridge: Cambridge University Press.
Henry, Meghan, Alvaro Cortes, and Sean Morris. 2013. "The 2013 Annual Homeless Assessment Report (AHAR) to Congress." Washington, DC: US Department of Housing and Urban Development; Office of Community Planning and Development.
Henig, Jeffrey R. 1994. *Rethinking School Choice: Limits of the Market Metaphor.* Princeton, NJ: Princeton University Press.
Henig, Jeffrey R., and Jason A. MacDonald. 2002. "Locational Decision of Charter Schools." *Social Science Quarterly* 83 (4): 962–980.
Hicklin, Aaron. 1995. *Boy Soldiers.* Edinburgh: Mainstream.
Higate, Paul R. 2003. "Introduction: Putting Men and the Military on the Agenda." In *Military Masculinities: Identity and the State,* edited by Paul R. Higate, xvii–xxii. Westport, CT: Praeger Publishers.

Higate, Paul, and John Hopton. 2004. "War, Militarism and Masculinities." In *The Handbook of Studies on Men and Masculinities*, edited by Michael S. Kimmel, Jeff Hearn, and R. W. Connell, 432–447. London: Sage Publications.

Hirsch, Barry T., and Stephen L. Mehay. 2003. "Evaluating the Labor Market Performance of Veterans Using a Matched Comparison Group Design." *Journal of Human Resources* 38 (3): 673–700.

Hirschfield, Paul. 2009. "School Surveillance in America: Disparate and Unequal." In *Schools under Surveillance: Cultures of Control in Public Education*, edited by Torin Monahan and Rodolfo Torres, 38–54. New Brunswick, NJ: Rutgers University Press.

Hoch, Paul. 1979. *White Hero, Black Beast: Racism, Sexism and the Mask of Masculinity*. London: Pluto Press.

Holton, Paul, Mark Bromely, Pieter D. Wezeman, and Siemon T. Wezeman. 2013. *Trends in International Arms Transfers, 2012*. Solna, Sweden: Stockholm International Peace Research Institute.

Hondagneu-Sotelo, Pierrette. 2001. *Domestica*. Berkeley, CA: University of California Press.

Houppert, Karen. 2005. "Military Recruiters Are Now Targeting Sixth Graders. Who's Next?" *The Nation*, September 12. http://www.thenation.com/article/whos-next.

Howard, John W. III, and Laura C. Prividera. 2004. "Rescuing Patriarchy or Saving 'Jessica Lynch': The rhetorical Construction of the American Woman." *Women & Language* 27 (2): 89–103.

Humensky, Jennifer L., Neil Jordan, Stroupe T. Kevin, and Denise M. Hynes. 2013."How Are Iraq/Afghanistan-Era Veterans Faring in the Labor Market?" *Armed Forces and Society* 39 (1): 158–183.

Jackson, Philip. 1968. *Life in Classrooms*. New York: Holt, Rinehart & Winston.

Janowitz, Morris. 1975. "The All-Volunteer Military as a 'Sociopolitical' Problem." *Social Problems* 22 (3): 432–449.

Jencks, Christopher, and Meredith Phillips. 1998. *The Black-White Test Score Gap*. Washington, DC: Brookings Institution.

Jennings, Chrstian, and Adrian Weale. 1996. *Green-Eyed Boys*. London: Harper Collins.

Jensen, Arthur R. 1980. *Bias in Mental Testing*. New York: Free Press.

Jones, Ann. 2013. "America's Child Soldiers: The Pentagon's JROTC Program Canvasses Public High Schools for Future Soldiers." *The Nation*, December 16. http://www.thenation.com/article/177603/americas-child- soldiers#.

Kantrowitz, Barbara, and Claudia Kalb. 1998. "Boys Will Be Boys." *Newsweek*, May 11, 55.

Karp, Stan. 2004. "NCLB's Selective Vision on Equality: Some Gaps Count More than Others." In *Many Children Left Behind: How the No Child Left Behind Act Is Damaging our Children and our Schools*, edited by Deborah Meier and George Wood, 53–65. Boston, MA: Beacon Press.

Kehily, Mary Jane. 2000. "Understanding Heterosexualities: Masculinities, Embodiment and Schooling." In *Genders and Sexualities in Educational*

Ethnography, edited by Geoffrey Walford and Caroline Hudson, 27–40. New York: Elsevier Science Inc.

Kemp, Janet, and Robert Bossarte. 2012. *Suicide Data Report, 2012*. Washington, DC: Department of Veteran Affairs, Mental Health Services Suicide Prevention Program. http://www.va.gov/opa/docs/suicide-data-report-2012-final.pdf.

Kessler, S., D. J. Ashenden, R. W. Connell, and G. W. Dowsett. 1985. "Gender Relations in Secondary Schooling." *Sociology of Education* 58 (1): 34–48.

Kimmel, Michael S. 1999. "'What about the Boys?' What the Current Debates Tell Us (and Don't Tell Us) about Boys in School." *Michigan Feminist Studies* 14: 1–28.

———. 2003 [1994]. "Masculinity as Homophobia." In *Reconstructing Gender: A Multicultural Anthology*, edited by Estelle Disch, 103–109. Boston, MA: McGraw Hill.

Kirk, Gwyn. 2008. "Contesting Militarization: Global Perspectives." In *Security Disarmed: Critical Perspectives on Gender, Race, and Militarization*, edited by Barbara Sutton, Sandra Morgen, and Julie Nokov, 30–55. Piscataway, NJ: Rutgers University Press.

Klein, Naomi. 2007. *The Shock Doctrine: The Rise of Disaster Capitalism*. New York, NY: Picador.

Kleinfield, Judith. 1999. "Student Performance: Males Versus Females." *Public Interest* 134: 3–20.

Klemm Analysis Group. 2000. *Evaluation Summary Montgomery GI Bill Program*. Washington, DC: Department of Veterans Affairs. www.va.gov/OPP/eval/MGIB_Eval.pdf.

Kleykamp, Meredith A. 2007. "Military Service as a Labor Market Outcome." *Race, Gender & Class* 14 (3/4): 65–75.

Knickerbocker, Brad. 1999. "Young and Male in America: It's Hard Being a Boy." *Christian Science Monitor*, April 29, 1.

Kozol, Jonathan. 1991. *Savage Inequalities: Children in America's Schools*. New York: Crown Publishers.

———. 2005. *The Shame of the Nation: The Restoration of Apartheid Schooling in America*. New York: Crown Publishers.

Kupchik, Aaron, and Torin Monahan. 2006. "The New American School: Preparation for Post-industrial Discipline." *British Journal of Sociology of Education* 27 (5): 617–631.

Ladson-Billings, Gloria. 2006. "From the Achievement Gap to the Education Debt: Understanding Achievement in US Schools." *Educational Researcher* 35 (7): 3–12.

Larabee, David F. 1997. "Public Goods, Private Goods: The American Struggle over Educational Goals." *American Educational Research Journal* 34 (1): 39–81.

Lasser, Jon, and Deborah Tharinger. 2003. "Visibility Management in School and Beyond: A Qualitative Study of Gay, Lesbian, Bisexual Youth." *Journal of Adolescence* 26 (2): 233–244.

Layton, Lyndsey. 2013. "U.S. Student Lag Around Average on International Science, Math and Reading Test." *The Washington Post*, December 2. http:

//www.washingtonpost.com/local/education/us-students-lag-around-average-on-international-science-math-and-reading-test/2013/12/02/2e510f26-5b92-11e3-a49b-90a0e156254b_story.html#.

Lehman, Hilary. 2011. "Army: NASCAR Pairing Paying Off." *Daytona Beach News*, February 20. http://www.newsjournalonline.com//racing/racing-business/2011/02/20/army-Nascar-pairing-paying-off.html.

Leonardo, Zeus, and W. Norton Grubb. 2014. *Education and Racism*. New York, NY: Routledge.

Letts, Will. 2006. "I Can't Even Think Straight: Queer Childhood and Adolescence." *Journal of Gay & Lesbian Issues in Education* 3 (2–3): 163–166.

———. 1998. "Educational Vouchers: Effectiveness, Choice and Costs." *Journal of Policy Analysis and Management* 17 (3): 373–392.

Levy, Yagil. 1998. "Militarizing Inequality: A Conceptual Framework." *Theory and Society* 2 (6): 873–904.

Lipman, Pauline. 2004. *High Stakes Education: Inequality, Globalization, and Urban School Reform*. New York: Routledge.

———. 2011a. "Cracking Down: Chicago School Policy and the Regulation of Black and Latino Youth." In *Education as Enforcement: The Militarization and Corporatization of Schools*, 2nd ed, edited by Kenneth J. Saltman and David A. Gabbard, 73–91. New York: RoutledgeFalmer.

———. 2011b. "Neoliberal Education Restructuring: Dangers and Opportunities of the Present Crisis." *Monthly Review* 63 (3): 114–127.

Lipman, Pauline, and David Hursch. 2007. "Renaissance 2010: The Reassertion of Ruling Class Power through Neoliberal Policies in Chicago." *Policy Futures in Education* 5 (2): 160–178.

Logan, John R., Elisabeta Minca, and Sinem Adar. 2012. "The Geography of Inequality: Why Separate Means Unequal in American Public Schools." *Sociology of Education* 85 (3): 287–301.

Losen, Daniel J., and Russel J. Skiba. 2010. *Suspended Education: Urban Middle Schools in Crisis*. Montgomery, AL: Southern Poverty Law Center. http://escholarship.org/uc/item/8fh0s5dv.

Loveless, Tom. 2003. "Charter School Achievement and Accountability." In *No Child Left Behind?: The Politics and Practice of School Accountability*, edited by Paul E. Peterson and Martin R. West, 177–196. Washington, DC: Brookings Institution.

Lubienski, Christopher. 2004. "Charter School Innovation in Theory and Practice: Autonomy, R&D, and Curricular Conformity." In *Taking Account of Charter Schools*, edited by Paul T. Hill, Katrina E. Bulkley, and Priscilla Wohlstetter, 72–90. New York, NY: Teachers College Press.

Lucas, Samuel Roundfield. 1999. *Tracking Inequality: Stratification and Mobility in American High Schools*. New York: Teachers College Press.

Lutz, Catherine. 2002. "Making War at Home in the United States: Militarization and the Current Crisis." *American Anthropologist* 104 (3): 723–35.

Lutz, Catherine, and Lesley Bartlett. 1995. "JROTC: Making Soldiers in Public Schools." *Education Digest* 61 (3): 9–15.

Mac an Ghaill, Mairtin. 1996. "What about the Boys? – Schooling, Class and Crisis Masculinity." *Sociological Review* 44 (3): 381–397.

MacLeod, Jay. 1987. *Ain't no Makin' It.* Boulder, CO: Westview Press, Inc.

Macmillan, Lorraine. 2011. "Militarized Children and Sovereign Power." In *The Militarization of Childhood*, edited by Beier J. Marshall, 61–76. New York: Palgrave Macmillan.

Majors, Richard. 2001. "Cool Pose: Black Masculinity and Sports." In *The Masculinities Reader*, edited by Stephen Whitehead and Frank Barrett, 208–217. Cambridge, MA: Polity Press.

Mann, Michael. 1992. *States, War and Capitalism: Studies in Political Sociology.* Oxford, UK: Basil Blackwell.

Manno, Bruno V., Gregg Vanourek, and Chester E. Finn, Jr. 1999. "Charter Schools: Serving Disadvantaged Youth." *Education and Urban Society* 31 (4): 429–445.

Matthews, Dylan. 2012. "Defense Spending in the U.S., in Four Charts." *The Washington Post*, August 28. http://www.washingtonpost.com/blogs/wonkblog/wp/2012/08/28/defense-spending-in-the-u-s-in-four-charts/.

McDuffee, Allen. 2008. "No JROTC Left Behind: Are Military Schools Recruitment Pools?" *In These Times*, August 20. http://www.inthesetimes.com/article/3855/.

McLaren, Peter. 1999. *Schooling as a Ritual Performance: Toward a Political Economy of Educational Symbols and Gestures.* Lanham, MD: Rowman & Littlefield.

———. 1998. *Life in Schools.* New York: Longman.

McMichael, Philip. 2008. *Development and Social Change.* Thousand Oaks, CA: Pine Forge Press.

McRobbie, Angela. 1978. "Working-class Girls and the Culture of Femininity." In *Women Take Issue: Aspects of Women's Subordination*, edited by Women's Studies Group. Hutchinson, London: University of Birmingham.

———. 1984. "Dance and Social Fantasy." In *Gender and Education*, edited by Angela McRobbie and Mica Nava, 130–161. London: Macmillian Publishers, Ltd.

Medina, J. 2007. "Recruitment by Military in Schools Is Criticized." *The New York Times*, September 7. http://query.nytimes.com/gst/fullpage.html?res=9E03E3DD133AF934A3575AC0A9619C8B63.

Miceli, Melinda 2010. "In the Trenches: LGBT Students Struggle with School and Sexual Identity." In *Sex Matters: The Sexuality and Society Reader*, edited by Mindy Stombler, Dawn M. Baunach, Elisabeth O. Burgess, Denise J. Donnelly, Wendy O. Simonds, and Elroi J. Windsor, 185–193. New York: Pearson.

Mills, C. Wright. 1956. *The Power Elite.* New York: Oxford University Press.

Miller, Toby. 2001. *Sportsex.* Philadelphia, PA: Temple University Press.

Milner IV, Richard H. 2013. *Policy Reforms and De-Professionalization of Teaching.* Boulder, CO: University of Colorado, National Education Policy Center. http://nepc.colorado.edu/publication/policy-reforms-deprofessionalization

Miron, G., J. L. Urschel, W. J. Mathis, and E. Tornquist. 2010. *Schools without Diversity: Education Management Organizations, Charter Schools, and* the *Demographic Stratification of the American School System.* Boulder, CO: Education and the Public Interest Center and Education Policy Research Unit.

Monzo, Lilia D. 2005. "Latino Parents' Choice for Bilingual Education in an Urban California School: Language Politics in the Aftermath of Proposition 227." *Bilingual Research Journal* 29 (2): 365–386.

Monzo, Lilia D., and Robert Rueda. 2009. "Passing for English Fluent: Latino Immigrant Children Masking Language Proficiency." *Anthropology & Education Quarterly* 40 (1): 20–40.

Morest, Vanessa Smith. 2002. "Privatization and Racial Segregation: A National Study of the Expansion and Institutionalization of the Charter School Movement." PhD diss., Department of Education, Columbia University, Ann Arbor, MI.

Morgan, David H. J. 1994. "Theater of War: Combat, the Military, and Masculinities." In *Theorizing Masculinities*, edited by Harry Brod and Michael Kaufman, 165–182. Thousand Oaks, CA: Sage Publications.

Moser, Michele, and Ross Rubenstein. 2002. "The Equality of Public School District Funding in the United States: A National Status Report." *Public Administration Review* 62 (1): 63–72.

Moskos, Charles. 2005. "A New Concept of the Citizen-Soldier." *Orbis* 49 (4): 663–676.

Nantais, C., and M. Lee. 1999. "Women in the United States Military: Protectors or Protected? The Case of Prisoner of War Melissa Rathbun-Nealy." *Journal of Gender Studies* 8 (2): 181–191.

Navarro, Vincent. 2006. "The Worldwide Class Struggle." *Monthly Review* 58 (4): 18–33.

Nazario, Sonia. 2007. "Junior ROTC Takes a Hit in LA." *Los Angeles Times,* February 19, A1.

Neas, Ralph G. 2003. *Funding a Movement: U.S. Department of Education Pours Millions into Groups Advocating School Vouchers and Education Privatization.* Washington, DC: People for the American Way.

Neill, Monty, Lisa Guisbond, Bob Schaeffer, James Madison, and Life Legeros. 2004. *Failing Our Children – How 'No Child Left Behind' Undermines Quality and Equity in Education: An Accountability Model that Supports School Improvement.* Cambridge, MA: The National Center for Fair and Open Testing. http://epsl.asu.edu/epru/articles/EPRU-0405-62-OWI.pdf.

Neilsen, Joyce McCarl, Glenda Walden, and Charlotte A. Kunkel. 2000. "Gendered Heteronormativity: Empirical Illustrations in Everyday Life." *Sociological Quarterly* 41 (2): 283–296.

Nelson, Beryl, Paul Berman. John Ericson, Nancy Kamprath, Rebecca Perry, Debi Silverman, Debra Solomon. 2000. *The State of Charter Schools 2000: National Study of Charter Schools Fourth-year Report.* Washington, DC: US Department of Education.

Nelson, Howard F. et al. 2004. *Charter School Achievement on the 2003 National Assessment of Educational Progress: Executive Summary.* Washington, DC: American Federation of Teachers. www.aft.org/pubs-reports/downloads/teachers/NAEPCharterSchoolReport.pdf.

Noonan, Margaret E., and Christopher J. Mumola. 2007. *Veterans in State and Federal Prison, 2004*. Washington, DC: US Department of Justice, Bureau of Justice Statistics. http://www.bjs.gov/content/pub/pdf/vsfp04.pdf.

Orenstein, Peggy. 1995. *Schoolgirls: Young Women, Self-esteem and the Confidence Gap*. New York: Random House.

Ozga, Jenny, H. Busher, and R. Saran. 1995. "Deskilling a Profession: Professionalism, Deprofessionalisation and the New Managerialism." In *Managing Teachers as Professionals in Schools*, edited by Hugh Busher and Rene Sara, 21–37. London: Kogan Page.

Paechter, Carrie. 2006. "Masculine Femininities/feminine Masculinities: Power, Identities and Gender." *Gender and Education* 18 (3): 253–263.

Paige, Rod, and Donald H. Rumsfeld. 2002. "Joint Letter from Secretary Paige and Secretary Rumsfeld." http://www.ed.gov/policy/gen/guid/fpco/hottopics/ht10-09-02c.html.

Parker, Andrew. 1996. "The Construction of Masculinity within Boys' Physical Education." *Gender and Education* 8 (2): 141–157.

Pascoe, C. J. 2007. *Dude, You're a Fag: Masculinity and Sexuality in High School*. Los Angeles, CA: University of California Press.

Passel, Jeffrey S., Randy Capps, and Michael Fix. 2004. *Undocumented Immigrants: Facts and Figures*. Urban Institute Immigration Studies Program. Washington, DC: Urban Institute.

Pedroni, Thomas C. 2007. *Market Movements: African American Involvement in School Voucher Reform*. New York: Routledge.

Perl, Libby. 2013. *Veterans and Homelessness*. Washington, DC: Congressional Research Service. Cornell University ILR School. http://digitalcommons.ilr.cornell.edu/key_workplace/1197/?utm_source=digitalcommons.ilr.cornell.edu%2Fkey_workplace%2F1197&utm_medium=PDF&utm_ca mpaign=PDFCoverPages.

Peterson, Paul E. 2003. "Introduction: After Zelman v. Simmons-Harris, What Next?" In *The Future of School Choice*, edited by Paul E. Peterson, 1–24. Stanford, CA: Hoover Institutional Press Publication.

Pipher, Mary. 1994. *Reviving Ophelia: Saving the Selves of Adolescent Girls*. New York: Penguin Group.

Plummer, Ken. 1996. "Symbolic Interactionism and the Forms of Homosexuality." In *Queer Theory/Sociology*, edited by Steven Seidman, 64–82. Cambridge, MA: Blackwell Publishers.

Raffaele-Mendez, Linda M., and Howard M. Knoff. 2003. "Who Gets Suspended from School and Why: A Demographic Analysis of Schools and Disciplinary Infractions in a Large School District." *Education and Treatment of Children* 26 (1): 30–51.

Rasmussen, Mary Louise. 2004. "Safety and Subversion: The Production of Sexualities and Genders in School Spaces." In *Youth and Sexualities*, edited by Mary Louise Rasmussen, Eric Rofes, and Susan Talburt, 131–152. New York: Palgrave Macmillan.

Renold, Emma. 2000. "'Coming Out': Gender, (Hetero)sexuality and the Primary School." *Gender and Education* 12 (3): 309–326.

Reynolds, C.R., R. J. Skiba, S. Graham, P. Sheras, J. C. Conoley, and E. Garcia-Vazquez. 2008. "Are Zero Tolerance Policies Effective in the Schools?: An Evidentiary Review and Recommendations." *American Psychologist* 63 (9): 852–862.

Rice, Jennifer King. 2003. *Teacher Quality: Understanding the Effectiveness of Teacher Attributes*. Washington, DC: Economic Policy Institute.

Rimer, Sara. 2004. "Unruly Students Facing Arrest, Not Detention." *New York Times*, January 4, A1.

Robbins, Christopher G. 2008a. "'Emergency!' Or How to Learn to Live with Neoliberal Globalization." *Policy Futures in Education* 6 (3): 331–350.

———. 2008b. *Expelling Hope: The Assault on Youth and the Militarization of Schooling*. Albany, NY: SUNY Press.

Robinson, Kerry H. 2005. "'Queerying' Gender: Heteronormativity in Early Childhood Education." *Australian Journal of Early Childhood* 30 (2): 19–28.

Rofes, Eric, and Lisa M. Stulberg. 2004. *The Emancipatory Promise of Charter Schools: Toward a Progressive Politics of School Choice*. Albany, NY: SUNY Press.

Rourke, Matt. 2007. "Veterans Make up 1 in 4 Homeless." *USA Today*, November 7. http://www.usatoday.com/news/nation/2007-11-07-homeless-veterans_N.htm.

Ross, Marlon B. 1998. "In Search of Black Men's Masculinities." *Feminist Studies* 24 (3): 599–626.

Sadker Myra, and David Sadker. 1994. *Failing at Fairness: How America's Schools Cheat Girls*. New York: Scribner.

Saltman, Kenneth J. 2000. *Collateral Damage: Corporatizing Public Schools: A Threat to Democracy*. Oxford, UK: Rowland & Littlefield Publishers.

———. 2003. "Introduction." In *Education as Enforcement: The Militarization and Corporatization of Schools*, edited by Kenneth J. Saltman and David A. Gabbard, 1–24. New York: RoutledgeRalmer.

———. 2007. *Capitalizing on Disaster: Taking and Breaking Public Schools*. Boulder, CO: Paradigm Publishers.

Saltman, Kenneth J., and David A. Gabbard (eds). 2003. *Education as Enforcement: The Militarization and Corporatization of Schools*. New York: RoutledgeRalmer.

———. 2010. *The Gift of Education: Public Education and Venture Philanthropy*. New York: Palgrave Macmillan.

——— (eds). 2011. *Education as Enforcement: The Militarization and Corporatization of Schools*. 2nd ed. New York: RoutledgeRalmer.

———. 2012. *The Failure of Corporate School Reform*. Boulder, CO: Paradigm Publishers.

Savage, Charlie. 2004. "Military Recruiters Target Schools Strategically." *The Boston Globe*, November 29. http://www.boston.com/news/nation/articles/2004/11/29/military_recruiters_pursu e_target_schools_carefully/.

Schaeffer-Duffy, Claire. 2003. "Uncle Sam Hustles to Keep the Ranks Filled". *COCO in the News*, March 21. http://www.objector.org/ccco/inthenews/sam_hustles.html.

Schott Foundation. 2004. *Public Education and Black Male Students*. Cambridge: Schott Foundation.

Schneider, Mark, and Jack Buckley. 2002. "What Do Parents Want from Schools? Evidence from the Internet." *Educational Evaluation and Policy Analysis* 24 (2): 133–144.

Schur, Edwin M. 1984. *Labeling Women Deviant: Gender, Stigma and Social Control.* New York: Random House.

Schwartz, Pepper, and Virginia Rutter. 1998. *The Gender of Sexuality: Exploring Sexual Possibilities.* Thousand Oaks, CA: Pine Forge Press.

Seidman, Steven. 1996. "Introduction." In *Queer Theory/Sociology*, edited by Steven Seidman, 1–29. Cambridge, MA: Blackwell Publishers.

Selman, Donna and Paul Leighton. 2010. *Punishment for Sale: Private Prisons, Big Business, and the Incarceration Binge.* Annapolis, MD: Rowman & Littlefield Publishing Groupo, Inc.

Sewell, Tony. 1997. *Black Masculinities and Schooling: How Black Boys Survive Modern Schooling.* Stoke-on-Tent, UK: Trentham Books.

Simon, Jonathan. 2006. *Governing through Crime: How the War on Crime Transformed American Democracy and Created a Culture of Fear.* New York: Oxford University Press.

Simon, Stephanie. 2012. "Private firms Eyeing Profits from U.S. Public Schools." *Reuters*, August 2. http://www.reuters.com/article/2012/08/02/usa-education-investment- idUSL2E8J15FR20120802.

Singer, Peter Warren. 2011. *Corporate Warriors: The rise of the Privatized Military Industry.* Ithaca, NY: Cornell University Press.

SIPRI. 2013a. *SIPRI Arms Transfers Database, 2013.* Solna, Sweden: Stockholm International Peace Research Institute. http://portal.sipri.org/publications/pages/transfer/tiv-data.

———. 2013b. *SIPRI Military Expenditure Database, 2013.* Solna, Sweden: Stockholm International Peace Research Institute. http://www.sipri.org/research/armaments/milex/milex_database.

Skelton, Christine. 2001. *Schooling the Boys: Masculinities and Primary Education.* Philadelphia, PA: Open University Press.

Skiba, R., C. R. Reynolds, S. Graham, P. Sheras, J. C. Conoley, and E. Garcia-Vazquez. 2006. *Are Zero Tolerance Policies Effective in the Schools? An Evidentiary Review and Recommendations.* Washington, DC: American Psychological Association Zero Tolerance Task Force.

Smith, Andrea. 2005. *Conquest: Sexual Violence and American Indian Genocide.* Cambridge, MA: South End Press.

Smith, George W. 1998. "The Ideology of 'Fag': The Fostering of Maleness in One Primary School." *Sociological Quarterly* 39 (2): 309–335.

Sontag, Susan. 1964 [1999]. "Notes on Camp." In *Camp: Queer Aesthetics and the Performing Subject: A Reader*, edited by Fabio Cieto, 53–65. Edinburgh, UK: Edinburgh University Press.

STARBASE. 2012. *2012 STARBASE Annual Report.* Washington, DC: Department of Defense.

———. *STARBASE 2.0: Program Overview/Fact sheet.* Washington, DC: Department of Defense.

Statistical Abstract of the United States. 2006. *Table 501*. Washington, DC: US Census Bureau.

Stein, Arlene, and Ken Plummer. 1996. "'I Can't Even Think Straight': 'Queer' Theory and the Missing Sexual Revolution in Sociology." In *Queer Theory/Sociology*, edited by Steven Seidman, 129–144. Cambridge, MA: Blackwell Publishers.

Steinhauer, Jennifer. 2013. "Reports of Military Sexual Assault Rise Sharply." *New York Times*, November 7. http://www.nytimes.com/2013/11/07/us/reports-of-military-sexual-assault-rise-sharply.html.

Stephen, Lynn. 2008. "*Los Nuevos Desaparecidos y Muertos*: Immigration, Militarization, Death, and Disappearance on Mexico's Borders." In *Security Disarmed: Critical Perspectives on Gender, Race, and Militarization*, edited by B. Sutton, S. Morgen, and J. Novkov, 79–100. New Brunswick, NJ: Rutgers University Press.

Stiehm, Judith Hicks. 1982. "The Protected, The Protector, The Defender." In *Women and Men's Wars*, edited by Judith Hicks Stiehm, 367–376. Oxford: Paragon.

Stulberg, Lisa M. 2008. *Race, Schools, and Hope: African Americans and School Choice after Brown*. New York, NY: Teachers College Press.

Sutton, Barbara, and Julie Novkov. 2008. "Rethinking Security, Confronting Inequality." In *Security Disarmed: Critical Perspectives on Gender, Race, and Militarization*, edited by Barbara Sutton, Sandra Morgen, and Julie Novkov, 3–29. New Brunswick, NJ: Rutgers University Press.

Tannen, Michael B. 1987. "Is the Army College Fund Meeting Its Objectives" *Industrial and Labor Relations Review* 41 (1): 50–62.

Taylor, Howard F. 1981. "Biases in Bias in Mental Testing." *Contemporary Sociology* 10: 172–174.

———. 2002. "Deconstructing the Bell Curve: Racism, Classism, and Intelligence in America." In *2001 Race Odyssey: African Americans and Sociology*, edited by Bruce R. Hare, 60–76. Syracuse, NY: Syracuse University Press.

———. 2003. "Inequality and the Bell Curve: Analyzing the Heritability and Race-Gender Bias of Cognitive Test Scores." Colloquium paper presented at Department of Sociology, Princeton University, Princeton, NJ.

Taylor Jr, William J. 1999. *Junior Reserve Officers' Training Corps: Contributing to America's Communities*. Washington, DC: The CSIS Press.

Thorne, Barrie. 1993. *Gender Play: Girls and Boys in School*. New Brunswick, NJ: Rutgers University Press.

Thurlow, Crispin. 2002. "Naming the 'Outsider Within': Homophobic Pejoratives and the Verbal Abuse of Lesbian, Gay, and Bisexaul High-School Pupils." *Journal of Adolescence* 24 (1): 25–38.

Troops to Teachers. 2014. "Troops To Teachers, Zone 1: Overview." http://troopstoteachers.net/Zones/Zone1/AbouttheProgram/Overview.aspx

Trudell, Bonnie Nelson. 1993. *Doing Sex Education: Gender Politics and Schooling*. New York: Routledge.

Turner, Ralph H. 1960. "Sponsored and Contest Mobility and the School System." *American Sociological Review* 25 (6): 855–867.

Turpin, Jennifer. 1998. "Many Faces: Women Confronting War." In *The Women and War Reader*, edited by L. Lorentzen and J. Turpin, 3–18. New York: New York University Press.

United States Army Recruiting Command. 2004. *School Recruiting Program Handbook*. Fort Knox, KY: Headquarters, US Army Recruiting Command.

US Census. 2014. *State and County Quick Facts: California*. Washington, DC: US Census Bureau. http://quickfacts.census.gov/qfd/states/06000.html.

US Department of Education. 1965. *Elementary and Secondary Education: Title I, Part A*. Washington, DC: US Department of Education.

———. 1983. *A Nation at Risk: The Imperative for Educational Reform*. National Commission on Excellence in Education. Washington, DC: US Department of Education.

———. 2008. "Troops-to-Teachers Program." http://www.ed.gov/programs/troops/index.html.

———. 2013a. "Table 205 Summary of Expenditures for Public Elementary and Secondary Education, by Purpose: Selected years, 1919–20 through 2009–10." Washington, DC: National Center for Educational Statistics, Common Core of Data. http://nces.ed.gov/programs/digest/d12/tables/dt12_205.asp.

———. 2013b. "Table 216.30 Number and percentage distribution of public elementary and secondary students and schools, by traditional or charter school status and selected characteristics: Selected years, 1999–2000 through 2011–12." Washington, DC: National Center for Educational Statistics, Common Core of Data. http://nces.ed.gov/programs/digest/d13/tables/dt13_216.30.asp.

———. 2013c. "Table 216.90 Public elementary and secondary charter schools and enrollment, by state: Selected years, Selected years, 1999–2000 through 2011–12." Washington, DC: National Center for Educational Statistics, Common Core of Data. http://nces.ed.gov/programs/digest/d13/tables/dt13_216.90.asp.

Valadez, James R., and Richard P. Durán. 2007. "Redefining the Digital Divide: Beyond Access to Computers and the Internet." *The High School Journal* 90 (3): 31–44.

Venetis, Penny et al. 2011. "Targeting Youth: What Everyone Should Know about Military Recruiting, Military Life, and Veteran Affairs Before Enlisting." A report prepared by the Constitutional Litigation Clinic, Rutgers School of Law-Newark. http://law.newark.rutgers.edu/files/u/MilitaryRecruitingReportConLitFinal.pdf.

Vine, David. 2009. "Proposals for a Humanpolitik: Building a New Human-Centered Foreign Policy." In *The Counter-Counterinsurgency Manual: Or, Notes on Demilitarizing American Society*, edited by Network of Concerned Anthropologists Steering Committee, 173–190. Chicago: Prickly Paradigm Press.

Volante, Louis. 2004. "Teaching To the Test: What Every Educator and Policymaker Should Know." *Canadian Journal of Educational Administration and Policy* 35. http://www.umanitoba.ca/publications/cjeap/articles/volante.html.

Wald, Joanna, and Daniel J. Losen. 2003. "Defining and Re-directing a School-to-Prison Pipeline." In *New Directions for Youth Development: Deconstructing* the *School- to-Prison Pipeline*, edited by J. Wald and D. Losen, 9–16. Hoboken, NJ: Wiley & Sons.

Waldner-Haugrud, Lisa K., and Brian Magruder. 1996. "Homosexual Identity Expression among Lesbian and Gay Adolescents: An Analysis of Perceived Structural Associations." *Youth and Society* 27 (3): 313–333.

Walford, Geoffrey. 2000. "Introduction." In *Gender and Sexualities in Educational Ethnography*, edited by Geoffery Walford and Caroline Hudson, 1–6. Amsterdam, the Netherlands: JAI.

Walker, James C. 1985. "Rebels with Our Applause? A Critique of Resistance Theory in Paul Willis's Ethnography of Schooling." *Journal of Education* 167 (2): 63–83.

Walters, Andrew S., and David M. Hayes. 1998. "Challenging the Culturally Sanctioned Dismissal of Gay Students and Colleagues." *Journal of Homosexuality* 35 (2): 123.

Warner, Michael. 1993. "Introduction." In *Fear of a Queer Planet: Queer Politics and Social Theory*, edited by Michael Warner, vii-xxxi. Minneapolis, MN: University of Minnesota Press.

Weaver-Hightower, Marcus. 2003. "The 'Boy Turn' in Research on Gender and Education." *American Educational Research Association* 73 (4): 471–498.

Wedekind, Jennifer. 2005. "The Children's Crusade: Military Programs Move into Middle Schools to Fish for Future Soldiers." *In These Times*, June 3. http://www.inthesetimes.com/article/2136/.

Weiher, Gregory R., and Kent L. Tedin. 2002. "Does Choice Lead to Racially Distinctive Schools? Charter Schools and Household Preferences." *Journal of Policy Analysis and Management* 21 (1): 79–92.

Wells, Amy Stuart, Julie Slayton, and Janelle Scott. 2002. "Defining Democracy in the Neoliberal Age: Charter School Reform and Educational Consumption." *American Educational Research Journal* 39 (2): 337–361.

Weis, Lois, and Michelle Fine. 2001. "Extraordinary Conversations in Public Schools." *Qualitative Studies in Education* 14 (4): 497–523.

Weisman, Robert. 2006. "Colleges Craft Studies to Fit Defense Firms." *Boston Globe*, June 27, E1.

Welch, Kelly. 2007. "Black Criminal Stereotypes and Racial Profiling." *Journal of Contemporary Criminal Justice* 23 (3): 276–288.

Wells, Kristopher. 2006. "Learning to Transgress: Queer Young Adults, Emotional Resilience, and Intellectual Resistance as Impetus for Lifelong Learning for Social Justice." In the *Proceedings of the 47th Annual Adult Education Research Conference*, 458–464, University of Minnesota, MN.

West, Candace, and Don H. Zimmerman. 1987. "Doing Gender." *Gender and Society* 1 (2): 125–151.

White House. 2012. *Military Skills for America's Future: Leveraging Military Service and Experience to Put Veterans and Military Spouses Back to Work*. Washington, DC: Executive Office of the President.

Whitehead, John W. 2013. "The End of Childhood in the Era of the Emerging American Police State." *Huffintonpost.com*, December 17. http://www.huffingtonpost.com/john-w-whitehead/the-end-of-childhood-in- t_b_4453122.html.

Willis, Paul E. 1977. *Learning to Labor*. Aldershot, UK: Gower.
Wood, Julian. 1984. "Grouping towards Sexism: Boy's Sex Talk." In *Gender and Generation*, edited by Angela McRobbie and Mica Nava, 280-303. New York: Routledge.
Woodward, Rachel. 2000. "Warrior Heroes and Little Green Men: Soldiers, Military Training, and the Construction of Rural Masculinities." *Rural Sociology* 65 (4): 640-657.
Yancey, Patty. 2000. *Parents Founding Charter Schools: Dilemmas of Empowerment and Decentralization*. New York: Peter Lang.
Žižek, Slavoj. 2008. *Violence: Six Sideways Reflections*. New York: Picador.

Index

absolute poverty, 148
accountability
    see under neoliberal ideology
African Women's Peace Networks, 155
Althusser, Louis, 15, 16, 17
Amanda, 126
American Association of University Women (AAUW) report (1992), 98
American Friends Service Committee (AFSC), 150, 154
antiracist, 148, 150
Apple, Michael, 86
Armed Services Vocational Aptitude Battery (ASVAB), 42, 84, 152-3, 158n2
Army College Fund, 48, 49
"asking out," 125-6, 135

Bakhtin, Mikhail M., 120
battle dress uniforms (BDUs), 77, 89
Ben, 115, 125-6, 132, 133-4
Bernstein, Basil, 17, 18
bisexual identity/bisexuality, 112, 128, 132, 133, 134, 136
Black/African-Americans, 18, 50, 53, 61
    US military, 43, 48-9
Black boys at the MEI, 72
    discipline, 109, 113
    hegemonic masculinity, 32, 100, 115
    violence, 109, 113
    weapon knowledge, 107
Black girls at the MEI, 72, 133, 133
    hegemonic masculinity, 100, 115
    resistance to gendered heteronormativity, 133-4
    rumors about, 124
Black masculinity, 109
Bourdieu, Pierre, 16-17
Bourdieu and Passerson, 16
Bowles and Gintis, 15-16, 17, 76
Bridgette, 109-10, 112, 124, 125, 126, 127-8, 132, 133
Bryan, 125
Bulter, Judith, 117

Cadet Code of Conduct, 65, 73
California Cadet Corps, 25, 35
California Education Code: 57605(5A-P), 62
camouflage, 77, 89, 108, 112, 159n4
Candace, 133
cannon ceremony, 88-9, 90, 94, 95, 159n4
capitalism, 56, 143
    see also neoliberal ideology, neoliberalism
carnivalesque, 120, 121, 134, 136
charter schools, 2, 6, 7, 24, 63, 80
    achievement, 9, 21
    California, 23, 24, 61-2, 157n1
    defined, 8
    drawbacks to, 8-11
    militarized, 2, 9, 142
    New Orleans, 6
    segregation (race and class), 8-9, 21, 142-3
    teachers, 9

chasing games, 102, 111, 127–8, 135
Child Soldiers Prevention Act, 56, 84, 95, 150
choice
   see under neoliberal ideology
Chris, 73
Christina, 123, 125, 126
Chubb and Moe, 7
The Citadel, 25
Clarissa, 71
Claudia, 68, 71, 73, 75, 80
Coalition Against Militarism in Schools, 150
college
   see higher education
"college prep," 64, 65
combat readiness, 91
commandants at MEI, 1, 26, 28, 29, 30, 68, 77, 92, 106, 107, 157n4, 157n7
competition
   see under neoliberal ideology
Connell, R. W., 20, 88, 100–1, 109
contested space, 121, 128, 131, 145
Convention on the Rights of the Child (2002), 84
counter militarization, 154–5
counter-hegemonic sexual personas, 133, 134
counterrecruitment, 150, 153, 154

"daily dozen," 139
dancing, 112, 130–1, 133, 134, 136
"dangerous youth," 9, 44–5, 53, 64, 76, 81
Danny, 108
Deferred Action for Childhood Arrivals (DACA), 158n2
Delayed Entry Program, 84, 159n3
democracy
   public schools, 8, 85, 86, 151–2
   see also under neoliberalism, public education
democratic futures, 147
dirty dancing, 123, 130, 136
disaster capitalism, 5, 39

discipline
   Black boys at MEI, 109, 113
   Foucault, 7, 59, 75, 126
   MEI, 72–5, 76, 77, 81, 142
   militarized, 4, 36, 45, 50, 64, 81, 141
   youth of color, 44, 52–3, 141
   see also under neoliberalism
discourse, 86–7
discursive symbols, 87–90, 92
Doug, 102–3, 104, 107–8 dress blues, 77, 89, 128

Eastmoore community, 25–6, 62, 67, 68, 70, 79, 142
   militarization of, 25–6, 88–9, 90
Eastmoore Middle School, 23–4, 62–3, 69–71, 72, 73
Eastmoore School District, 23–4, 62–3, 69, 70, 77, 80–1, 142
   racial tension, 63, 70–1
   teacher experience and credential rates, 65–6
   test scores, 67
   violence, 63, 70–2, 77
education
   see also under public education, schools
efficiency
   see under neoliberal ideology
Eisenstein, Zillah, 3–4, 148
emergency teaching permits, 65–6, 158n3
Enloe, Cynthia, 3–4, 148
Equality of Education Opportunity Report (1966), 38
Estella, 71, 75
ethnographic studies, 19, 20, 22, 116, 133, 144
expulsions
   MEI, 64, 74, 109, 110
   schools, 53, 98

Federal Development, Relief and Education for Alien Minors (DREAM) Act, 50–1, 53, 57
female masculinity. *See* masculine girls

femininity, 102, 111–13, 127
  "girlie-girls," 112–13
  MEI, 98, 111–13, 127, 128
  militarization, 37
  warrior hero, 102
feminism, 4, 98
feminist analysis, 148–9, 150
feminist organizations, 150, 155
fieldtrips, 27, 90, 91, 92–3, 95, 106
flag ceremony, 88–9, 95
flirting/flirting scripts, 122, 125–8, 129
  same-sex, 125–8, 135–6
football, 108
foreign policy, 154
Foucault, Michel, 19, 86, 115, 117, 120
  discipline, 7, 59, 75, 126
freedom
  See under neoliberalism
Friedman, Milton, 5, 7, 14, 42, 55–6

Gabbard, David A., 4
gangs, 37, 71, 73, 78–9
"gay," 132
gay bashing, 133
gender, 37, 105, 100, 102–3, 105, 148
  schools, 20, 98–9, 116, 117–18, 119
gender resistance, 110, 111, 112–13
gender/sexuality regimes, 20, 117–18, 119, 122–3, 125, 127
  resistance to, 119–21, 124, 125–8, 129–31, 133–4
GI Bill. *See* Montgomery GI Bill
"girlie-girls," 112–13
girls, 98, 111, 124, 126–8, 130
Giroux, Henry, 18, 41, 53, 55, 86
Global Action to Prevent War and Conflict, 155
global capitalism, 143
Gramscian consent, 60–1, 87
Gun-Free Schools Act (1994), 52
guns
  airgun, 110
  masking tape gun, 106–7, 108
  rifles, 1, 87, 106

rockets, 107–8, 110
and women, 104

Halberstam, J., 111
Heath, Shirley Brice, 18
Hector, 107–8
hegemonic masculinity, 97–8, 100, 101, 105, 109, 110
heteronormalizing practices, 132
heteronormative gendered practices, 121–4, 125, 127, 130–1
heteronormative practices, 116, 119, 121, 122–3, 125, 127–9, 130–1
heterosexual matrix, 117–18
heterosexual practices, 122–4, 127, 130
heterotopia/heterotopic space, 120, 121, 124, 126, 128, 129, 131
"hidden curriculum," 75–6
higher education
  MEI cadets, 70, 94
  militarization, 43
  veterans, 47, 49
Hispanic Access Initiative, 43
homophobia, 132–3
  masculinity, 99, 131–2
  teasing, 133
How Schools Shortchange Girls, 98
Hurricane Katrina, 6

Illegal Immigration Reform and Immigrant Responsibility Act (1996), 50
Incite! Women of Color Against Violence, 150
International Women's Network against Militarism, 150, 155

Jackie, 132
JoAnn, 109–10, 112, 124, 127–8, 132, 133
Joint Advertising and Market Research Studies (JAMRS), 153
Jones, Mr., 23, 63, 73, 91, 130, 139
Junior Reserve Officer Training Corps (JROTC), 44–7, 84
  at-risk students, 45–6
  critiques, 46–7

Junior Reserve Officer Training
    Corps—*Continued*
  girls, 44
  higher education, 46
  resistance, 153–4

Klein, Naomi, 5
knives, 92–3

language
  demilitarize, 149
  MEI, 144, 149
Latino/as, 9, 20, 22, 43, 50
  US military, 22, 43, 48–9
Lawrence, 78
lesbians, 110, 112, 132, 133
Linda, 132

Maricela, 102, 111–12, 127
Marilis, 71
masculine girls, 100, 109–10, 111, 124
masculinity
  female, 109–10, 100, 111–12, 113
  hegemonic, 97–8, 100, 101, 105, 109, 110
  homophobia, 99, 131–2
  marginal, 100, 109
  militarized, 37, 88, 99, 101–2, 105, 148
  race, 37, 100, 109, 113
  schools, 20, 98–9, 118
  subordinated, 100
  violence, 101, 105
MEI mascot, 87–8
MEI motto, 90–1, 139
Middle School Cadet Corps, 84, 158n1
Miles, 70
militarism, 85–6, 89, 90, 95
  defined, 4
  compare militarization, 3–4
militarization, 36
  defined, 3–4
  gender, 4, 37
  neoliberalism, connection with, 2, 14, 40, 41–2, 53, 55, 85
  public education, 14, 41, 43, 44–56, 53, 85, 99

  and race, 4, 36
  resistance to, 149–55
  social status, 79
  STEM fields, 52
  US, 36
  youth, 36–7, 52, 55
  compare militarism, 3–4
militarized
  discipline, 4, 36, 45, 50, 64, 76, 81
  identities, 85, 88, 90, 92, 94
  language, 149
  language at the MEI, 90–2
  masculinity, 88, 99, 101–2, 105, 148
  nation state, 35–6, 45, 53, 55
  pedagogies, 85, 87, 89, 91, 92, 94
  schools, 25, 45, 68
  sexuality, 104–5
  violence, 108, 148, 149
military
  global expenditure, 35, 148
  privatization, 42
  schools, 45, 68
  see also US military
military ball, 123, 128–31
military drill, 29, 94, 112–13, 124, 127, 133
Military Educational Institute (MEI), 22–3, 24, 62–3
  computer access, 65
  curriculum, 65, 106
  enrollment, 23–4, 64–5, 69, 76
  friendship groups, 72, 103–4, 127, 133–4
  gender, 101–5, 109–10, 111–13
  hierarchical structure of, 92
  militarization, 87, 90, 92, 94, 96, 144
  militarized discipline at, 73–5, 76, 81
  physical structure of, 64, 126–7
  race, 24, 72, 76, 109, 110–11, 113, 124, 133–4
  racial tension, 70–1, 81
  sexualities, 122–8, 129–31, 133–4
  social class, 24, 72, 76, 77
  social status, 59, 79–80, 89

Mills, C. Wright, 36
Montgomery GI Bill, 47-9
morning formation, 1, 73, 91, 139

A Nation at Risk (1983), 39, 56, 140
National Commission on Excellence in Education, 39
National Defense Authorization Act (2006), 45
National Network Opposing Militarization of Youth (NNOMY), 150
National Priorities Project, 151
neoliberal ideology, 6, 7, 38
  accountability, 7, 11, 12
  choice, 7, 10-11, 63
  competition, 7, 10
  efficiency, 7, 12, 70
neoliberal nation state, 6, 7, 40-1, 53, 54, 55, 77, 81
neoliberalism
  challenges to, 147-8, 150, 154
  defined, 2
  democracy, 6, 40, 85, 86, 147-8
  discipline, 7, 8, 9, 14, 41, 52-3, 76
  freedom, 6, 40, 60-1, 147
  globalization, 41, 55
  militarization, connections with, 2, 14, 40, 41-2, 53, 55, 85
  moments of crisis, 5-6, 39
  pedagogy, 40, 61, 85-6, 150
  privatization, 6, 11, 12, 13, 80
  public education, 5-6, 7, 14, 39, 61, 140
  social inequalities, 6-7, 40-1, 143, 148
No Child Left Behind Act (NCLB), 10, 11-12, 13, 43, 84
  critiques, 12-13
  military recruitment, 43, 83-4, 152
nonviolence, 149

Oi, Walter, 42
"operations,"
  see fieldtrips
opt-out waivers, 83, 152

parents, 60, 61, 83, 152
MEI, 29, 63, 69-70, 72-3, 75, 76, 78, 79-80, 109, 123, 130, 131
  militarization, 92
  school choice, 60
Park, Mr., 70, 72, 75, 76, 78
peace education, 152, 154
pedagogies
  critical, 85-6, 151
  militarized, 85, 87, 89, 91, 92, 94, 95, 150
  neoliberal, 40, 61, 85-6
Peter, 107-9, 110
pleasure, 120, 130, 131
Project on Youth and Non-military Opportunities, 150
protector
  see warrior hero
public displays of affection, 122-4, 125
  heterosexual, 122-4
  same-sex, 124
public education
  choice, 10-11
  class, 13, 15-18, 19, 61, 147
  crisis in, 5, 39, 56, 140
  democracy, 8, 85, 86, 151-2
  funding, 8, 151-2
  history of, 37-9
  inequalities, 12, 38-9, 61
  militarization, 14, 41, 43, 44-56, 53, 85, 99
  militarized schools, 25, 45, 68
  neoliberalism, 5-6, 7, 14, 39, 61, 140
  privatization, 5, 6, 12, 13
  race, 13, 18, 20, 21, 38, 61, 80
  racism, 61, 143, 147
  sites of resistance, 151-4
  see also schools

queer cadets, 112, 124, 128, 130, 132, 134
queer theory, 116-17, 118-19
queer youth, 118, 119, 120

188 INDEX

Race to the Top program (RTT), 11, 13–14
racism, 143, 147, 148, 150
  discourses about youth, 44
  MEI, 109
  school punishment, 53, 109
recruitment
  see under US military
Reserve Officers' Training Corps (ROTC), 43
resistance at MEI
  femininity, 128
  gendered heteronormativity, 124, 127, 128, 129, 131, 133, 134
  homophobia, 133
  masculinity, 112–13, 124
  military culture, 112–13
resistance theory, 18–19, 144
resistance to militarization
  community level, 149–50
  global level, 154–5
  institutional level, 150–4
  personal level, 149
rifles, 1, 87, 106
Robbins, Christopher G., 53, 64
rockets, 107–8, 110
ROTC Vitalization Act (1964), 44
Rough Rider Ration, 90
The Rough Riders, 87–8
ruling class, 61, 143
rumors, 124, 132–3, 134

Saltman, Kenneth, 4, 6, 151
Sam, 109, 110
same-sex desire, 125, 126–7, 128, 131
San Diego V. Governing Board of Grossmont Union High School District [790 F.2d 1471 (9th Cir, 1986)], 153
Sarah, 1, 103, 108
school boards, 8, 151–2
The School Recruiting Program Handbook, 84
schools
  class, 15–18, 19
  gender, 20, 98–9

gender/sexuality regimes, 20, 117–18, 119
homophobia, 132
  as an institution, 15, 85, 117–18, 119, 121
race, 18, 19, 20
sites of resistance, 119–20, 151
see also charter schools, public education
Science, Technology, Engineering, and Math (STEM), 51–2
science projects, 107–8, 110
Servicemen's Readjustment Act, 47
sexual education, 122
sexual identities, 119
sexual scripts, 122, 125, 126, 127, 128, 129, 158n3
  flirting, 122, 125–8, 129
  military ball, 123, 128–31
  public displays of affection, 122–4, 125
Shawn, 106–7, 108
smash and grab capitalism, 6, 39
social class
  MEI, 24, 72, 76, 77
  public education, 13, 15–18, 19, 61, 147
  schools, 15–18, 19
  US military, 48, 49, 140, 141, 142
social reproduction theory, 15–18, 116, 144
The Solomon Amendment (1996), 43
stalking, 126–7
standardized testing, 7, 11, 12–13, 66, 158n2
  MEI, 66–8, 69
STARBASE/STARBASE 2.0, 51–2
"Stay Alert! Stay Alive!," 90–1, 139
Stockholm International Peace Research Institute, 151
Strong, First Sergeant, 23, 90, 92
subaltern counter publics, 120–1, 124, 125–6
survival, 91, 92–3, 106
Susan Up pal, 71, 72, 75, 131

INDEX

"talking shit," 134
teachers
    charter schools, 9
    credentials and experience at the MEI, 65–6
    deprofessionalization, 50
    militarization, 49–50
    teasing, 78, 79, 111, 127, 132, 133
"that's so gay," 132
theoretical implications, 3, 21–2, 105, 116–17, 118, 144, 145–6
theories of education
    resistance theory, 18–19, 144
    social reproduction theory, 15–18, 116, 144
Thomas, Mr., 59, 70, 78, 79, 80
Thorne, Barrie, 20
Title I, 12, 23, 62, 158n1
"to go out," 125–6
tomboyism, 111–12, 127
tough girls, 111, 127, 128
Troops to Teachers (TTT), 49–50, 84, 99
    race, 50

Uncle Sam, 158n1
undocumented students, 50–1, 158n3
uniforms, 65, 68, 76, 77–80, 88, 89, 111, 112–13, 128, 129–30, 131
Universal Declaration of Human Rights, 150
US imperialism, 57, 148
US military
    advertisements, 43–4, 158n1
    ban on gays and lesbians, 132
    budgets, 35, 42, 44, 45, 83, 148, 150
    employment, 36, 83
    personnel, 36
    privatization, 42
    race, 22, 32, 54
    recruitment, 42–4, 45–6, 55, 56, 83–4, 95, 140
    social class, 48, 49, 140, 141, 142
    women, 22, 24, 43–4, 103–5, 157n2

veterans
    domestic prisoners, 55
    education, 47, 49, 55
    employment, 49, 55
    homelessness, 54
    suicide, 54
Veteran's Educational Assistance Program (VEAP), 47
violence
    Eastmoore School District, 63, 64, 70–2
    legitimate, 37, 45, 88
    masculinity, 101, 105
    militarized, 108, 148, 149
    schools, in, 44–5, 52
    symbolic, 16
    systemic, 63
    youth, 53, 64
violence at the MEI
    acts of, 74, 107, 108–9
    condonement of, 108–9, 110, 111
    culture of, 91, 105, 107, 109, 111
    gender, 109, 110
    race, 109, 110
war, 35, 143
War Resister's League, 151
warrior hero, 88, 101–3
    girls at the MEI, 102
    masculinity, 88, 101–2
    women, 102–4
weapons, 52, 103, 105
    Black cadets, 106–7
    female cadets, 106, 110
    knives, 92–3
    knowledge of, 105–7, 108, 110, 111
    rockets, 107–8, 110
    see also guns
West, Major (commandant), 26–7, 28, 30, 68–9, 88, 90, 91, 106, 159n4
West Point, 25, 73
Willis, Paul, 18, 19
Wilson, Commandant, 28, 30, 77
Winchester Rifle logo, 87–8

women, 22, 24, 30, 43–4, 102–3, 103–5
   antimilitarization organizations, 150, 155
   guns, 104
   peace work, 149–50
   US military, 22, 24, 43–4, 103–5, 157n2
   warrior hero, 102–4
Women for Genuine Security, 150
Women of Color Resource Center (UC Berkeley), 150
Women's International League for Peace and Freedom, 155
wrestling, 111, 112, 127–8

Youth Attitudinal Tracking Survey, 42
YouTube.com, 44
Yssenia, 123

Zack, 70
zero-tolerance policies, 41, 44, 52–3, 109
Žižek, Slavoj, 63

GPSR Compliance
The European Union's (EU) General Product Safety Regulation (GPSR) is a set
of rules that requires consumer products to be safe and our obligations to
ensure this.

If you have any concerns about our products, you can contact us on

ProductSafety@springernature.com

In case Publisher is established outside the EU, the EU authorized
representative is:

Springer Nature Customer Service Center GmbH
Europaplatz 3
69115 Heidelberg, Germany

www.ingramcontent.com/pod-product-compliance
Lightning Source LLC
LaVergne TN
LVHW051912060526
838200LV00004B/99